INTERNATIONAL SERIES IN
NATURAL PHILOSOPHY
General Editor: D. ter HAAR

VOLUME 106

WAYNFLETE LECTURES ON PHYSICS

Other Pergamon Titles of Interest

Books

DEMIANSKI
Relativistic Astrophysics

GINZBURG
Theoretical Physics and Astrophysics

GURZADYAN
Flare Stars

KRAUSE and RADLER
Mean-field Magnetohydrodynamics and Dynamo Theory

*Journals**

Chinese Astronomy and Astrophysics

Journal of Quantitative Spectroscopy and Radiative Transfer

Planetary Space and Science

Plasma Physics

Vistas in Astronomy

*Free specimen copy available on request

A full list of titles in the International Series on Natural Philosophy follows the index

WAYNFLETE LECTURES ON PHYSICS

Selected Topics in Contemporary Physics and Astrophysics

by

V. L. GINZBURG

P. N. Lebedev Institute of Physics of the Academy of Sciences of the USSR,
Moscow, USSR

Translated by
D. ter Haar

PERGAMON PRESS

OXFORD · NEW YORK · TORONTO · SYDNEY · PARIS · FRANKFURT

U.K.	Pergamon Press Ltd., Headington Hill Hall, Oxford OX3 0BW, England
U.S.A.	Pergamon Press Inc., Maxwell House, Fairview Park, Elmsford, New York 10523, U.S.A.
CANADA	Pergamon Press Canada Ltd., Suite 104, 150 Consumers Road, Willowdale, Ontario M2J 1P9, Canada
AUSTRALIA	Pergamon Press (Aust.) Pty. Ltd., P.O. Box 544, Potts Point, N.S.W. 2011, Australia
FRANCE	Pergamon Press SARL, 24 rue des Ecoles, 75240 Paris, Cedex 05, France
FEDERAL REPUBLIC OF GERMANY	Pergamon Press GmbH, Hammerweg 6, D-6242 Kronberg-Taunus, Federal Republic of Germany

---

First edition 1983

Library of Congress Cataloging in Publication Data

Ginzburg, V. L. (Vitalii Lazarevich), 1916-
Waynflete lectures on physics.
(International series in natural philosophy; v. 106)
Includes index.
1. Physics. 2. Astrophysics. I. Title. II. Series.
QC21.2.G57 1983 530 82-24619

British Library Cataloguing in Publication Data

Ginzburg, V. L.
Waynflete lectures on physics.—(International series in natural philosophy v. 106)
1. Physics
I. Title II. Series
530 QC21.2

ISBN 0-08-029147-3

In order to make this volume available as economically and as rapidly as possible the author's typescript has been reproduced in its original form. This method unfortunately has its typographical limitations but it is hoped that they in no way distract the reader.

Printed in Great Britain by A. Wheaton & Co. Ltd., Exeter

Preface

In February 1982 I gave, as the 1982 Waynflete Lecturer, a series of seven lectures in Oxford. The present booklet contains the text of these lectures — which had been prepared beforehand in Russia. As one might expect, the lectures themselves differed slightly from and were shorter than the published text, but only on inessential points. The first four lectures dealt with the material of Chapters 1 and 2, the fifth lecture with that of Chapter 3, and the last two with that of Chapter 4. In writing the text I have on occasion used material which had been prepared before. I hope that the publication of the present booklet will be justified and will serve a useful purpose.

In conclusion I want cordially to thank the President and Fellows of Magdalen College, Oxford, for their kind invitation to deliver the Waynflete Lectures. I also want to thank Dirk ter Haar warmly, for his extensive help and for translating the text of the lectures into English.

Oxford, February 1982 V.L. GINZBURG

Contents

CHAPTER 1

Key Problems in Physics and Astrophysics

I. INTRODUCTION AND LIST OF PROBLEMS

I know from experience that selecting certain problems as being 'particularly important and interesting' or being 'key problems' sometimes generates strong objections, but more often will lead to a certain amount of misunderstanding and confusion. Therefore, before we start, it is necessary to clarify my position and state a number of reservations.

If we compare physics to an edifice, it would be a huge building with very strong foundations, with basement rooms, with spectacular towers, but also with access roads leading to factories, to hospitals, and to proving grounds. Thousands and thousands of people, hundreds of different installations and devices, thousands of problems and puzzles, hundreds of journals — all that can be found in the same place. There are many interconnections and there are many things which are difficult to separate or distinguish. On the other hand, separation and differentiation occur — for many years people work side by side, but they remain strangers. The building begins to resemble the Tower of Babel. At the same time, it is characteristic for physics — one should perhaps more correctly say, it is still characteristic for physics — that there exists a profound intrinsic unity of ideas and methods. The growth of the structure only makes it more difficult to recognize this unity and to trace and use it. Of course, at the level of the teaching of basic physics, including theoretical physics, there are no special problems; those appear in the final courses which lead to narrow specialization. One often meets the following situation: an excellent, able student or a research worker, author of original papers, has a very narrow horizon. He is familiar with the difficult, profound problems of quantum field theory, but he has no clue, not even at a qualitative level, about neutron stars, black holes, ferroelectrics, liquid ^{3}He and many other topics. At the same time, I am convinced of the following: Firstly, in order to be successful in one's work, it is very important to have a wide horizon, and, of course, it is also very interesting to be familiar with many topics. Secondly, it is not at all difficult to get acquainted, albeit only at a qualitative level, with a rather large field of problems, but one should take a little trouble and one should get some help.

These and similar considerations which one can easily present and develop, prompted me to write down a list of 'particularly important and interesting problems' in physics and astrophysics, and to organize a lecture course for students, and to write an article for the review journal Uspekhi Fizicheskikh Nauk. This article, entitled "which problems in physics and astrophysics are now particularly important and interesting" was published in 1971.* I pursued in the main a pedagogical goal, and I considered the composition of my list and the comments on it as a device or a means to widen the horizon of young physicists. However, up to the present I do not know the opinion of the youth. On the other hand, this paper produced totally unexpectedly, perhaps not a storm, but a deluge of reproaches from middle-aged and older physicist colleagues. Some of them felt that it is completely impossible to distinguish "particularly interesting and important problems", and others were offended because the list of such problems did not contain those in which they themselves were involved, and so on and so forth. In reply to this criticism I turned the article into a booklet, which has by now been translated into many languages.† This indicates at least that the discussion of 'key problems' causes interest. Recently I published a paper with the same title as in 1971, but with the subtitle "ten years on".≠ Its aim was not to compete with Dumas-père, but to trace the development of physics and astrophysics in the decade 1971 — 1981 using as an example a certain selection of key problems.

---

* Usp. Fiz. Nauk. **103** (1971) 87; Sov. Phys. Uspekhi, **14** (1971) 21.

† The most recent (third) edition of 'Physics and Astrophysics' appeared in Moscow in 1980. The English translation of this edition, including a number of additions is in the course of publication at Pergamon Press. In this booklet as well as in a recent paper quoted in the next footnote, one can find a large number of references to the literature about nearly all the problems which are touched upon in these lectures; this is the reason why I give hardly any references in this chapter.

≠ Usp. Fiz. Nauk, **134** (1981) 469; Sov. Phys. Uspekhi, **24** (1981) 585.

However, it is well known that the proof of the pudding is in the eating and it is now time to give finally the list which I have mentioned:

MACROPHYSICS

1. Controlled thermonuclear fusion.
2. High-temperature superconductivity.
3. New materials (the problem of producing metallic hydrogen and several other 'unusual' substances).
4. The metallic exciton (electron-hole) liquid in semi-conductors. Metal-dielectric transitions. Spin glasses.
5. Second-order phase transitions and phase transitions which are close to second-order ones (critical phenomena) with a number of interesting examples.
6. Surface physics.
7. The behaviour of matter in very strong magnetic fields.
8. The study of very large molecules. Liquid crystals.
9. X-ray and γ-ray analogues of lasers and new kinds of lasers.
10. Solitons. Strange attractors. Nonlinear phenomena.
11. Super-heavy elements (far transuranic elements). 'Exotic' nuclei.

MICROPHYSICS

12. Quarks and gluons. Quantum chromodynamics.
13. Unified theory of weak and electromagnetic interactions. $W^{-,0}$ bosons. Leptons.
14. Grand unification. Proton decay. Neutrino mass. Super-unification.
15. Fundamental length. Particle interactions at high and very high energies.
16. Breaking of CP-invariance. Nonlinear phenomena in vacuo in very strong magnetic and electric fields. Phase transitions in vacuo.

ASTROPHYSICS

17. Experimental verification and limits of applicability of the general theory of relativity.
18. Gravitational waves.
19. The cosmological problem. Connections between cosmology and high-energy physics.
20. Neutron stars and pulsars. Black holes and 'black hole' physics.
21. Quasars and galactic nuclei. Formation of galaxies.
22. The origin of cosmic rays and of the cosmic γ- and X-ray emission.
23. Neutrino astronomy.

The division of these problems into macrophysics, microphysics, and astrophysics is rather arbitrary. The selection of 23 of them and not, say, 20 or 30 is also arbitrary. The almost complete absence of problems connected with biology is due not to an underestimating of the importance of the relationship between physics and biology (quite the reverse: I have the greatest respect for and interest in biophysics), but to an attempt to restrict the list somehow or other and also to the fact that I feel incompetent to deal with biology. Finally, a major point: one cannot and should not think that one should deal only with 'key' problems. The latter are somehow distinguished and attract an exceptionally large amount of attention (this is, of course, reflected in the specialized and in the popular literature) by virtue of their potential technological importance, special mysteries, the existence of some nearly unsurmountable obstacles for finding solutions, and so on and so forth. On the other hand, however carefully one defines the concept of what are 'key problems' they cannot really stay key problems for ever and ever; moreover, there are scarcely even two physicists who would draw up completely identical lists of key problems. In the booklet which I referred to earlier these provisos have been dealt with in detail. In that booklet and also in a number of papers I asked my colleagues for constructive criticism — to draw up their 'lists' and comments — and for a discussion of the essential points. In this I have not been successful, but why this is so is not really clear to me.

One can give a course of lectures on each of the 23 problems I have listed, and certainly, at any rate a whole lecture. However, this is not possible for me in this series of lectures — and this is perhaps just as well. One should really ask a specialist in the topic discussed to give such a lecture — and we organize it in this way in the Physico-Technical Institute in Moscow, where I occupy the chair of physical and astrophysical problems. In the present set of lectures I devote separate lectures to only two of the 'key problems': to high-temperature superconductivity and to the origin of cosmic rays — to be more precise, to high-energy astrophysics which includes cosmic ray astrophysics. Two lectures will be devoted to emission by sources moving with a constant velocity: Cherenkov radiation, the Doppler effect, transition radiation, and transition scattering. There, one is mainly dealing with problems in the classical electrodynamics of continuous media. There are in that case no special mysteries or any other of the above-mentioned features which would entitle them to be included amongst the key problems. However, I have studied this kind of problem right from the start of my scientific activity and, daring to use a strong term, I love these problems. Their inclusion in this set of lectures is determined not only by the fact that they turn out to be interesting and instructive, but also in order to emphasize that what I have said earlier, namely, that there are no grounds whatever to restrict oneself to 'key problems'.

The second part of my first lecture — and also the next two lectures — will be devoted to commenting on the list of key problems which I have just given. Of course, I can only give fragmentary remarks and estimates, but it is clearly impossible to limit oneself to just giving a simple list — one should touch more definitely upon the scientific problems involved. The numbering of the subsections will correspond to the numbering of the problems in the list.

II. MACROPHYSICS

1. *Controlled thermonuclear fusion.* To solve the problem of controlled thermonuclear fusion means to learn how to use efficiently for energy purposes the reactions (d, t, p, and n stand, respectively, for the deuteron, the tritium nucleus, the proton, and the neutron):

$$d+d \to {}^3\mathrm{He}+n+3.27\ \mathrm{MeV}\ ,\qquad d+d \to t+p+4.03\ \mathrm{MeV}\ , \tag{1}$$

$$d+t \to {}^4\mathrm{He}+n+17.6\ \mathrm{MeV}\ . \tag{2}$$

When one uses tritium, as was assumed in the first stage, the reaction

$${}^6\mathrm{Li}+n \to t+{}^4\mathrm{He}+4.6\ \mathrm{MeV} \tag{3}$$

also plays an important part; it enables us to reproduce tritium which is radioactive and hardly occurs at all in nature. It may turn out that one could also use some other reactions, for instance, the reaction

$$d+{}^3\mathrm{He} \to {}^4\mathrm{He}+p+18.34\ \mathrm{MeV}\ .$$

Even in the case of the most effective reaction (2), the mixture of deuterium and tritium must be heated to a temperature $T \sim 10^8\ \mathrm{K} \approx 10^4$ eV, and one must satisfy the condition $n\tau > 3\times 10^{14}\ \mathrm{cm}^{-3}\ \mathrm{s}$, where n is the electron density in the plasma (which is, of course, completely ionized) and τ is the characteristic time during which the energy is contained in this plasma.

The basic difficulty for solving the problem — if, for the moment, we do not talk about inertial type devices — is connected with the confinement of a hot plasma. The evolution of the appropriate ideas is rather typical and instructive. At the beginning of this research, thirty years ago, when the problem had already been formulated rather clearly, it appeared that it would not be all that difficult to solve it (I myself was then involved in this work). But soon it became clear that a cavalry attack would not reach the goal. A hot plasma turned out to be a much more complicated and less well studied object than it seemed to be on the basis of earlier results, which had been accumulated from a study of the Earth's ionosphere and of gas discharges. It has already taken three decades for plasma physics to reach a certain degree of maturity: this process does not only enrich physics in general, but also turns out to be very important for understanding various processes in cosmic plasmas. It has been possible only through the thorny path of trial and error to understand how important, for instance, it is to control the perfection of the magnetic field — it was the fact that this was not realized which led at a certain stage to the disillusion with stellarators — and to obtain a practically total absence from the plasma of impurities (nitrogen, oxygen, and so on, even apart from nuclei with larger atomic numbers Z).

At the present time tokamak type devices have especially been further advanced. Their development, using superconducting coils, promises to lead to the desired goal by the end of this century. However, in as far as I can judge from data in the literature, stellarators and open systems (mirror machines) can fully compete with tokamaks. However, much is still obscure even as far as tokamaks are concerned — let us mention, as an example, the question of how to heat the plasma.

This is even more the case when we talk of inertial type thermonuclear devices. These involve the relatively fast compression of solid or liquid pellets consisting of a mixture of deuterium and tritium — or of pure deuterium — through irradiation by lasers, or by electron or ion beams. The possibilities of inertial type devices have not been studied as well as those of devices involving magnetic confinement of the plasma. I am insufficiently acquainted with these problems, but in the case of a 'laser thermopile' huge difficulties are visible even with the naked eye. Moreover, we have the difficulty of the low efficiency of lasers and the relatively short life of laser materials. Further, when we use ion beams, and especially when we use electron beams, there arise new difficulties. As far as I understand the situation, the problem is, in general, still far from having been solved, even in principle.

It is thus certainly too early to assume that the thermonuclear problem is a technical, engineering problem. It remains a physical problem of great interest as far as principles are concerned and one can hardly doubt that it belongs to the 'key problems'. It is likely that this will remain to be the case for the next one or two decades.

2. *High-temperature superconductivity*. This problem will be the subject of a separate lecture. It will become clear that this is a typical physics problem in that the results of future investigations remain completely

uncertain. It is thus impossible to predict when it will be solved and even whether it will be solved at all in a positive sense — in the sense of obtaining relatively easily accessible high-temperature superconductors. If the outcome is positive, but also if it is negative (in passing we note that it is very difficult in this case to reach a negative conclusion, as it is in general difficult to prove the impossibility of the existence of a whole range of possibilities which at first sight can be thought of), the problem of high-temperature superconductivity vanishes from the list of 'key problems'. It is, however, probable that this stage is still far off. At any rate, that it is impossible to give some date for the solution of the problem does in no way prevent us from understanding that a physical problem exists and that it is important.

3. *New materials (metallic hydrogen)*. As far as new materials are concerned — we are dealing with materials which in some sense or other are exotic as otherwise one cannot speak of a physical problem — we shall consider only metallic hydrogen. Under normal conditions (atmospheric pressure) hydrogen consists of molecules, boils at $T_b = 20.3$ K and solidifies at $T_m = 14$ K. The density of solid hydrogen, which is a dielectric, is $\rho = 0.076\ \mathrm{g\,cm^{-3}}$. However, under sufficiently strong compression, when the atomic shells are destroyed, all substances, amongst them hydrogen, must become metals. A very rough estimate of the density at which metallization takes place can be obtained from the condition $\rho_{metal} \sim Ma_0^{-3}$, where $M = 1.67 \times 10^{-24}$ g is the mass of a hydrogen atom and $a_0 = \hbar^2/me^2 = 0.529 \times 10^{-8}$ cm the Bohr radius. This criterion leads to a strongly over-estimated density $\rho_{metal} \sim 10\ \mathrm{g\,cm^{-3}}$, whereas more exact calculations lead to $\rho_{metal} \sim 1\ \mathrm{g\,cm^{-3}}$. For this one needs a pressure which so far has not yet been evaluated exactly but which is likely to be of the order $P_{metal} \sim 2$ to 3 Mbar. Metallic hydrogen, possibly metastable, may exist also at lower pressures (and, maybe even at atmospheric pressure). There are grounds for assuming that metallic hydrogen is a high-temperature superconductor with a critical temperature $T_c \sim 100$ to 200 K. A large part of the giant planets — Jupiter and Saturn — consists of metallic hydrogen (with impurities).

In general, there is no doubt about the exceptional importance and interest of the problem of metallic hydrogen. So much was clear long ago. However, it has only recently been realized how exceptionally hard it is to solve this problem. It is not difficult to produce a pressure of 3 M bar using shock waves, but, in general, heating occurs in that case and the compression phase is short-lived. Static pressures in the megabar range are also very well accessible; this can be done by using a 'pocket' type device fitting on an ordinary table. Of course, the compressed volume is then small — the name of the game is force per unit area (so that enormous pressures can exist at the point of a needle). However, to produce the necessary pressure is still unlikely. The main difficulty is the production of an 'anvil' which can stand up to the necessary pressure. Even diamond — the best of the materials known for such a purpose — starts to 'flow' at pressures above 1.7 to 2 M bar. As a result no reliable, certain metallization of hydrogen has been accomplished. It is hard to predict when a 'piece' of metallic hydrogen will be obtained, but probably it will not be soon. One needs here new ideas. Let me repeat, it is clear that we are dealing here with a physical problem and not with a technological one.

The production of 'essentially new' (exotic) materials such as metallic hydrogen, anomalous water, high temperature superconductors, and so on, just like the search for new particles and nuclei (say, far transuranic elements) and so on, is very alluring and sometimes causes a rush. As a result, one quite often finds in the literature reports about various 'discoveries' which are subsequently not confirmed. The problem about how to treat such cases attracts a certain attention in scientific circles, and it is a controversial problem. That is the reason why I think that it is not inappropriate if I express here my own opinion about this point. I am against too strong a condemnation of authors who publish erroneous papers — assuming, of course, that we are dealing with a publication made in good faith and not one made unscrupulously. Firstly, the publication harms most of all the authors and in the case of errors nobody suffers worse than themselves. Secondly, the publication enables one to ascertain the truth faster, leading to checks by other people. This is necessary. It is absolutely impossible to agree with the requirement of some authors to have their results acknowledged before they are comprehensively checked. Within reasonable limits authors have the right to make errors, but all their colleagues have, at least, equal rights to have doubts.

4. *Metallic exciton liquid in semiconductors.* If there are electrons and holes (produced, say, by illumination) in a semiconductor, at a sufficiently low temperature they can be effectively combined into hydrogen-like 'atoms' — the excitons which are analogous to positronium. The binding energy $E_{0,e}$ and the characteristic size $a_{0,e}$ of such excitons in the ground state are

$$E_{0,e} \sim \frac{m_{eff}e^4}{2\hbar^2\varepsilon^2} = \frac{m_{eff}E_0}{m\varepsilon^2} \quad , \quad a_{0,e} \sim \frac{\hbar^2\varepsilon}{m_{eff}e^2} = \frac{a_0\varepsilon m}{m_{eff}} \quad . \tag{4}$$

Here $E_0 = e^4m/2\hbar^2$ and $a_0 = \hbar^2/e^2m$ are the well known expressions for the binding energy and radius of the hydrogen atom in its ground state, e and m are the charge and mass of a free electrons, and $\hbar = h/2\pi$ with h Planck's constant. Further, in (4) ε is the dielectric permittivity of the semiconductor and m_{eff} is the electron and hole effective mass (for the sake of simplicity we assume their masses to be the same; we neglect anisotropy). In a number of cases $\varepsilon \sim 10$ and $m_{eff} \leqslant 0.1\, m$; it thus happens that

$$E_{0,e} \leqslant 10^{-2}\text{ eV} \sim 100\text{ K} \quad \text{and} \quad a_{0,e} \geqslant 10^{-6}\text{ cm}.$$

As a result of the weakening of the Coulomb attraction by a factor ε and the decreases in the effective mass, the exciton thus turns out to be very loosely bound and enormous in size as compared with a hydrogen atom or positronium.

Hence it follows at once that for an exciton gas or liquid in a semiconductor the criteria for high density and metallization, which were mentioned earlier in connection with metallic hydrogen, will be satisfied at relatively low densities. Indeed, the rough condition for metallization,

$$n_e a_{0,e}^3 \sim 1, \tag{5}$$

is for $a_{0,e} \sim 10^{-6}$ cm satisfied for exciton densities $n_e \sim 10^{18}\text{ cm}^{-3}$ which is easily attainable. It is thus possible to imitate high pressures and, in actual fact, to produce an exciton (electron-hole) metallic liquid in a semiconductor. What an interesting object! At the present time, as the result of about ten years of research the metallic exciton liquid (this is in fact a liquid which forms drops, and so on) has been studied already rather well, but only in a few semiconductors. At the same time it is, in principle, possible that in other kinds of semiconductors a dielectric exciton liquid can exist. There is still the problem of superconductivity and superfluidity of exciton liquids. In the case of two-dimensional (surface) and one-dimensional (filament) systems excitons can also exist; the criteria for them to be metallized or having high densities (collectivization) differ from (5) in that $a_{0,e}^3$ must be replaced by $a_{0,e}^2$ or $a_{0,e}$. One may thus expect further studies of exciton liquids in semiconductors and it is as yet too early to eliminate this problem from our list. However, I realize that to assume that only the problem which we have just discussed is a key problem in semiconductor physics would be incorrect. However, I could say only very little about other problems which have been widely discussed in the literature: disordered systems and spin glasses (which may be either metals or semiconductors), metal-dielectric phase transitions, and several other ones.

5. *Phase transitions and critical phenomena*. This problem is, indeed, huge, and many books have been devoted to it*, apart from an enormous number of

---

* See, for instance, S.K. Ma, Modern Theory of Critical Phenomena, Benjamin, New York, 1976.

papers. Here I can only make a few fragmentary remarks in this connection.

When one neglects fluctuations one can use the self-consistent (molecular) field theory which goes back to van der Waals, Weiss, and others, and which was later generalized by Landau to the case of many order parameters and the inclusion of anisotropy. This kind of theory has often been used and is often relatively well applicable, although this is far from always the case. Enormous effort has been expended on generalizing the theory or, to be more precise, on taking fluctuations into account which are particularly important near a number of critical points and second-order phase transition points. As a result much has been achieved, such as the introduction and evaluation of critical indices. Some physicists even feel that the theory is already in a satisfactory state or that it is completely impossible to do any more. Personally, I disagree with the first and do not believe in the second statement. For example, if we look at the well known and widely studied λ-transition in helium II it is clearly obvious that a micro-theory of the transition has not been developed to any great extent, not even for the homogeneous case. When we change to non-uniform conditions (channels, gaps, films) and flow (including vortex lines) there is practically no micro-theory at all and basically we can use only the quasi-phenomenological Ψ-theory.* Of course, the theory of liquids, in this case of helium II, meets with special difficulties, but also in this field there is a certain amount of progress. It is difficult to believe that it will never be possible to develop a sufficiently complete micro-theory of the λ-transition in helium.

In the general field of phase transitions there are still unsolved problems, both in the theory and from an experimental point of view. Moreover, the whole time one's attention is drawn to various new systems and cases. We may mention, for instance, transitions between phases with incommensurate parameters in ferro-electrics, ferro-magnetics, or other solids. One should pay special attention to the phase transitions in liquid ^{3}He which were discovered ten years ago. Although the experiments take place at temperatures below 2 to 3×10^{-3} K and, sometimes, under pressure, the progress in this field has been striking. However, I could not start to discuss this here, even if there were time. The situation in the case of liquid ^{3}He is so much more complicated than in the case of ^{4}He, with which I am acquainted, that I almost right from the start decided not to deal with ^{3}He. Not without reason is it said that "physics is the game of the young". Fortunately, this remark

* See V.L. Ginzburg and A.A. Sobyanin, Sov. Phys. Uspekhi, **19** (1976) 733; J. Low Temp. Phys., in course of publication.

must be taken *cum grano salis*: physics encompasses also the work of people of a more advanced age. The latter, I feel, are, in general, able to judge the achievements of others. I therefore take the liberty to express my opinion that the study of liquid ^{3}He is the most prominent achievement in the field of condensed state physics during the last decade and, in particular, that it deserves the Nobel Prize.

Recently attention has also been paid to the possibility of a superfluid transition in atomic hydrogen gas. If one could somehow produce such a gas, it would under normal conditions quickly change into a gas of molecular hydrogen H_2. However, at low temperatures, $T \leqslant 1$ K, a gas of atomic hydrogen in a vessel with its walls covered by superfluid helium would live several minutes. If we would also place this gas in a strong magnetic field, the stability of atomic hydrogen would be even further enhanced and, apparently, its recombination can become unimportant — the cause for this is well known: in the H_2-molecule the electron spins are anti-parallel to one another, but in a strong magnetic field the spins of all the electrons would be pointing in the same direction so that for the formation of an H_2-molecule one of the spins must flip over and this is not easily done. H-atoms with parallel spins in the ground state repel one another — it is true that there is a weak van der Waals attraction at large distances apart, but this is of little importance. Such a gas will thus even under atmospheric pressure not be liquified down to absolute zero. However, at some temperature the real Bose-gas of H-atoms must undergo a Bose-Einstein condensation and become superfluid. Altogether this is a rather exotic system and its importance can hardly be compared with that of liquid ^{3}He, but it is interesting. There are also quite a number of other interesting and important problems in the field of the physics of phase transitions. Apart from superconductors and the λ-transition in helium II, I have myself relatively recently been concerned with the problem of possible superfluidity of molecular hydrogen and the scattering of light in the vicinity of phase transition points.* It would be inappropriate to dwell here upon these cases. In general, of course, one must emphasize that the problem of phase transitions remains one of the main problems in physics.

6. *Surface physics*. Various processes and phenomena on surfaces are, of course, not at all a new kind of problem. However, in the last decade deep changes have taken place in this field. We have learned, at least in some cases, to obtain and control the purity and the state of the surface. There

---

* See V.L. Ginzburg, A.P. Levanyuk, and A.A. Sobyanin, Phys. Repts., **57** (1980) 151.

have appeared many effective methods for the analysis of surfaces, or they have been improved: low energy electron diffraction (LEED), angle-resolved photo-emission spectroscopy (ARPS), inelastic scattering of ions with energies of the order of 1 MeV, electron microscopy, the study of surface acoustic waves and surface polaritons, and so on. Already many important results have been obtained. I may mention, for instance, the reconstruction of surfaces — the appearance on the surface of a lattice with a period which differs from that in the bulk of the substance — for instance under well-defined conditions in Si on the 111 face the lattice parameter is seven times larger than in bulk Si. Surface ordering deserves a special mentioning: the occurrence of surface ferro- and anti-ferromagnetism, surface ferro-electricity, surface superconductivity, and surface superfluidity. Besides, even a simpler situation is impressive: in a dielectric (semiconductor) electron surface states may be only partly occupied, which means, that the dielectric in the bulk may have a metallic surface conductivity. Phase transitions in two- or quasi-two-dimensional systems are closely related to this class of problems. Altogether this is an extremely quickly increasing range of investigations.

7. *The behaviour of substances in very strong magnetic fields*. The Zeeman splitting of the levels of atomic hydrogen in a magnetic field is

$$E_{\mathrm{H}} \sim e\hbar H/mc \sim 10^{-8} H \,\mathrm{eV};$$

the magnetic field H is here measured in Oersted or Gauss, which are the same, as I use exclusively the cgs system of units. On the other hand, the characteristic distance between levels when there is no magnetic field, or when the magnetic field is weak, is determined by the Coulomb interaction between the electron and the proton and is equal to $E_{\mathrm{a}} \sim e^4 m/2\hbar \sim 10\,\mathrm{eV}$. Clearly $E_{\mathrm{H}} \ll E_{\mathrm{a}}$ as long as

$$H \ll H_{\mathrm{cr},1} = \frac{e^3 m^2}{c\hbar^3} = \left(\frac{e^2}{\hbar c}\right)^2 \left(\frac{mc}{e\hbar}\right) mc^2 \sim 3 \times 10^9 \ \mathrm{Oe}\,. \tag{6}$$

In very strong magnetic fields, when $H \gg H_{\mathrm{cr},1}$, the effect of the magnetic field will become dominant, and the atom will be like a needle elongated along the field, that is, it will not at all look like the normal hydrogen atom. For a heavy atom with atomic number Z the magnetic field will be very strong, when

$$H \gg H_{\mathrm{cr,Z}} = Z^3 H_{\mathrm{cr},1} \sim 3 \times 10^9 \, Z^3 \ \mathrm{Oe}\,. \tag{7}$$

So far it has been possible to produce in a laboratorium only fields with $H \lesssim 10^6$ Oe and the problem of very strong fields thus seemed to be academic. However, soon after the discovery of pulsars in 1967—1968 it became clear that magnetic neutron stars possessed magnetic fields $H \sim 10^{12}$ to 10^{13} Oe.

Incidentally, this conclusion was clear even before the discovery of pulsars: when a star with a high conductivity changes from a 'normal' star of radius r_0 and field H_0 to a neutron star with parameters r_n and H_n flux is conserved, $H_0 r_0^2 \sim H_n r_n^2$; with $r_0 \sim 10^{11}$ cm and $H_0 \sim 10^2$ Oe we get thus for a neutron star with $r_n \sim 10^6$ cm a field $H_n \sim 10^{12}$ Oe.

From what we have just said it is clear that matter in neutron stars or in their vicinity may be situated in very strong magnetic fields. This fact is important for the theory of pulsars, but we cannot dwell upon this here. However, it is difficult not to note even now that it is possible also under terrestrial conditions that the magnetic field may dominate the Coulomb field. For instance, in semiconductors the magnetic field will be very strong, for the excitons (in which an electron and a hole are bound together) which we discussed earlier, when (see (6))

$$H \gg H_{cr,semic} = H_{cr,1}\left(\frac{m_{eff}}{m\varepsilon}\right)^2 \sim 3\times10^9\left(\frac{m_{eff}}{m\varepsilon}\right)^2 \text{ Oe}. \quad (8)$$

When $m_{eff} \sim 0.1\,m$ and $\varepsilon \sim 10$, the field $H_{cr,semic} \sim 3\times10^5$ Oe, which can be reached in a laboratory.

I hope that even the little I have said is sufficient to make clear how interesting the problem of the behaviour of matter in very strong magnetic fields is.

As I shall not dwell in what follows upon problem number 16 in my list, I note here that very strong fields — in fact, exceeding the field (6), as we are now contemplating a field $H_c = m^2c^3/e\hbar = 4.4\times10^{13}$ Oe — changes not only the properties of matter, but also those of the vacuum. In such fields $H \gtrsim H_c \sim 10^{13}$ Oe, and sometimes even in somewhat lower fields, the vacuum behaves as a nonlinear medium. This fact was recognized already in 1934, almost half a century ago. However, only recently has the problem of very

---

* I cannot refrain from giving the rather striking example of how simple physical considerations help in the understanding and even in the prediction of results which are obtained by solving a very complicated problem. For instance, it turns out that the vacuum in a strong magnetic field behaves, as far as the propagation of weak electromagnetic waves is concerned, as a birefringent, but not a magneto-active medium. At the same time it is well known that an ordinary electron-ion plasma and other substances in a magnetic field become magneto-active — in particular, the polarization plane rotates in it: the Faraday effect takes place. This difference becomes at once clear, if we bear in mind that the nonlinear properties of the vacuum in a magnetic field are caused by the creation of virtual electron-positron pairs. Now, such a 'virtual plasma' must behave similarly to a real electron-positron plasma with equal electron and positron densities. Because electrons and positrons rotate in a magnetic field in the opposite sense, such an electron-positron plasma is clearly not magneto-active, but, of course, it is birefringent, as there is a preferred direction, namely, that of the external magnetic field.

strong fields in vacuo become a real one, in the sense that it has become possible to observe various effects.*

So far I have managed to touch upon only seven problems out of the 23 from my list. It is clear that to proceed any further in the same spirit within the framework of the present lectures is impossible. The following exposition will therefore be even more fragmentary than the one which has been given up to now.

11. *Super-heavy (far transuranic) elements; exotic nuclei.* This problem refers to the field of the physics of atomic nuclei, but I have included it in the section on macro-physics. Of course, there were no strict principles involved in my classification. All the same, I feel, perhaps anticipating the arguments somewhat, that nuclear physics should on the whole be relegated rather to macro- than to micro-physics in the way it is understood nowadays (*vide infra*). Indeed, in nuclear physics we are dealing basically with non-relativistic particles such as nucleons, electrons, and, sometimes, positrons. The number of nucleons in heavy nuclei is rather appreciable and in many respects a nucleus is akin to a liquid drop. In general, the resemblance to macro-physics and, in particular, to atomic physics, is altogether much stronger than to high-energy physics, which is a typical representative of micro-physics. However, I repeat that there is no special virtue in having a particular classification.

In my list I included only one problem, number 11, from the field of nuclear physics and that was rather arbitrary and quite possibly because I am not particularly competent in this field. The essence of this problem is the search and study of 'exotic', as yet unknown nuclei. This includes far transuranic nuclei, that is, nuclei with $Z \sim 110$ to 114 with an enhanced stability (this relative stability is not excluded, because of shell effects, when $Z \approx 114$, $N = A - Z \approx 184$). Also included are nuclei with an unusual shape, hadronic atoms, such as an atom consisting of a proton and an antiproton, or nuclei with an increased density as compared to the normal nuclear density $\rho_n = 3 \times 10^{14}\ \mathrm{g\ cm^{-3}}$.

I myself have followed with more interest only one direction of these searches, namely, the attempts to observe exotic nuclei in cosmic rays. To be precise, if some isotope of a nucleus with $Z \sim 114$ would have a long lifetime, with a half-life $T \gtrsim 10^7$ to 10^8 years, such nuclei, formed, say, as the result of supernova outbursts, would form part of the cosmic rays which reach the solar system. In photographic emulsions, raised, of course, to the edge or beyond the edge of the atmosphere, there are as yet no reliable

* See footnote on previous page.

observations of far transuranic nuclei. In 1980 there appeared a report of the observation of a track of a nucleus with $Z \geqslant 110$ in an olivine crystal of meteoritic origin. The interpretation is, however, insufficiently unique and, in any case, for a safe inference one should find several tracks. The problem therefore remains unresolved.

I must mention also another trend in the search for far-transuranic elements, attempts to synthesize them in accelerators, in the same way as has already been achieved for elements with atomic numbers up to 107.

I am confident that the inclusion of the problem of exotic nuclei in my list of key problems does not meet with great opposition.

III. MICROPHYSICS

First of all: what is microphysics? Without some kind of agreement or definition there is no answer to this question. The boundary between macro- and micro-physics is in no way fixed; it is a historical category. At one time, half a century ago one would assign all of atomic physics to microphysics. I have mentioned earlier that nowadays not only atomic, but even nuclear physics belongs rather to macrophysics. Without any qualifications high-energy physics and the problem of the structure of matter at the 'fundamental' level (quarks, gluons, leptons) belong now to the field of microphysics. In general, the subject matter of the problems in microphysics is rather obvious and, in particular, is reflected in our list under the section 'microphysics'.

I resolved not to discuss the essence and contents of microphysics in the present set of lectures. This is not so much a question of lack of time, but rather one of principle. We are dealing here with the most profound problems in physics and, indeed, in the natural sciences. Just now there occur events of exceptional importance in this field (*vide infra*). At the same time I myself have recently not been occupied directly with microphysics problems, and only have attempted to follow the events through the literature, and this mainly through review and popular articles. Under those circumstances it would from my part be somewhat disrespectful both vis-à-vis the problems and vis-à-vis my present audience to talk about key problems in microphysics; the situation was different when in my booklet mentioned in the Introduction I was more or less forced, albeit with reservations, to touch also on the essence of microphysics. However, as an interested observer and physicist I have a right to make a few remarks of a general nature in the form of discussion points about microphysics. This I indend to do now in the hope that such remarks are here appropriate.

1. The development of the natural sciences and, in particular, of physics has already proceeded for three centuries, on the whole with rather great steadiness, growing roughly according to an exponential law with a growth of 5 to 7% per year; we refer here, say, to the number of papers, of journals, and of scientists. On the other hand, when one considers separate trends of research and particular problems, it is impossible to speak about uniformity: periods of violent growth and changes alternate with lulls and even stagnation. One should add to this that one should distinguish objective and subjective — or psychological — rates of growth of science. For a 30 or even a 40 year old person something which happened 20 years ago is for him very far away, prehistoric in terms of his own activity in science. It is also a fact that during a period of 20 years many new things occur indeed. As a result, there is in scientific circles a wide-spread overestimate of the rate at which science grows. In actual fact the development of science is on the whole much slower than it appears to be. It is sufficient to remember that the special theory of relativity is more than 75 years old, the general theory of relativity is not less than 65 years old, and quantum mechanics more than 50 years old. Superconductivity was discovered in 1911, but one had to wait 46 years before there was an explanation. Cosmic rays were discovered in 1912. However, both superconductivity and cosmic rays still remain to be objects to which close attention is paid and which are studied intensely. Briefly, during a period of 10, 20, or even 30 or 50 years much happens in physics, but, more importantly, very many things do not change radically. It is true that one meets with exceptional periods. For microphysics, yes, for physics as a whole, one may consider the 5 or 7 year interval between 1925 and 1930 or 1932 to be, apparently, such a period. During that time quantum mechanics was created, the basis was laid for relativistic quantum mechanics in quantum electrodynamics and Dirac's theory, a large number of concrete results where obtained in the field of atomic physics, and the positron and neutron were discovered, in 1932. However, even in that period, and later on, various difficulties appeared in the theory which took years and decades to overcome.

The most prominent success of the theory in the forties was the development of renormalization methods; as a result of this quantum electrodynamics became, if not completely consistent and without contradictions, at any rate such that it could, in principle, answer any problem put before it and that it admitted experimental verification. However, in the theory of weak and strong interactions there were for many years no striking successes. That should not cause any surprise: it is extremely difficult to advance in the field of fundamental problems, there are no well beaten tracks, and — as we said earlier, two or three decades are not such a long period.

This discussion was only necessary for me to express the suggestion — without any pretence of being original — that microphysics in the past decade and evidently up to the present moment, has experienced an exceptional period. Indeed, as far as ideas about the structure of matter are concerned the transition from nucleons to quarks and gluons has taken place. In the field of strong interaction theories the investigators have turned from some rather unsuccessful attempts to quantum chromodynamics. The unification of weak and electromagnetic interactions has taken place. Very interesting attempts at a grand or even super-unification of all interactions, including the gravitational interaction are made. All this occurs, even if not on a completely new basis, at any rate on a rich, high-principled basis, including symmetries, especially, the generalized gauge symmetry, spontaneous symmetry breaking, and the study of nonlinear equations with a veritable superabundance of possibilities. It is simply impossible to compare the situation with that in the fifties and sixties.

Of course, far from all problems have been solved and many of them remain completely unexplained, but that is practically always the case. One of the most intriguing problems is the following: does the 'subdivision' of matter stop at quarks or should one accomplish a transition to proto- or sub-quarks (preons or so). Moreover, there are also other, not less acute problems. Somehow or other microphysics experiences a period of take-off and progress — as we said, it may even be an 'exceptional' period which is excessively fruitful and valuable. In this connection the distinguished and fundamental position of microphysics as the most profound and principled part of physics, its leading edge or front, is especially clear.

2. On the background of what we have just said a second remark which I want to make — or, to be more precise, which I want to repeat, as it is contained in my papers and in the booklet which were quoted earlier — may even cause some surprise. I believe that the place or position occupied by microphysics in the natural sciences and the whole life of the human society has now become different and appreciably less important than was the case in the first half of our century. This remark seems to me to be trivial and obvious. However, it is a fact that far from everybody agrees with it and that I have encountered severe criticism because of it. I must therefore explain what I have said — and this will, maybe, not be without interest for my audience.

Interest in science, being occupied with science, and devotion to it do not need any justification. Being occupied by science is for many people a vocation, the main one in their life, just as for other people art, music, and so on plays an analogous rôle. However, one does not normally ask a musician why he plays, or why he devotes his life to music. On the other hand, physicicists are often asked: why do you do this work, what use will your work have for industry, for agriculture, and so on. Of course, I am strongly in favour that it will bear fruit, whenever this is possible. However, such a requirement cannot and should not figure as the main motivation for scientific activity. To be precise, working in the field of microphysics is itself completely justified, it is fascinating and attractive, independent of whether or not one can see any practical applications. In that sense, microphysics yesterday, today, and, one must feel, tomorrow, was, is, and will be the leading front of physics and of the whole of the natural sciences. Actual problems change, of course, but the spirit, the atmosphere, and even the style of the investigations in the field of microphysics remain to a very large extent the same. I always remember what Einstein said, when he characterized this atmosphere as "the years of anxious searching in the dark, with their intense longing, their alternations of confidence and exhaustion and the final emergence into the light".

However, every coin has two sides. I feel that this also applies to science; it is not isolated from society and is connected with it by tight bonds. It is therefore natural that public resonance, one's position, and the general value of some scientific branches and fields are determined not just by internal scientific considerations or the logic of the particular science, but also by its effect on the outside world. It is just in this scheme of things and in this sense that I feel that during the first half of our century microphysics occupied in the natural sciences a very special place, the first place. This was the result of an interference between the contents of microphysics and their influence on other disciplines and on society as a whole. Indeed, atoms, and later atomic nuclei were the objects of microphysics. It was extraordinarily difficult to understand the behaviour of atoms and of the electrons in atoms — to do this it needed a real revolution: the creation of quantum mechanics. In other words, microphysics of that period enabled us to understand — and as a consequence in many respects to control — the structure of matter at a level which we meet with in everyday life. Hence the vast effect of the corresponding successful advances on the whole of the natural sciences, such as chemistry, biology, on technology, medicine, the whole of life, and on the development of human society, including, unfortunately, the threat of the application of terrible weapons.

Present-day microphysics with its quarks, its particles which live for only a tiny fraction of a second, and its all-pervading neutrino has passed through hot waters and, if I may be excused such comparisons, has emerged into beautiful, but severe cold water. To make a long story short, let me say that I feel that the position of microphysics at the present time is similar to that of astrophysics, excluding the study of the solar system. And this is a very remarkable position! As far as I am concerned, what I have said is, firstly, the statement of facts: it is sufficient to look at the physics literature, both journals and books, to see the steep decline in the fraction of microphysics research as compared to the past — one must, of course, take into account the change in the contents of microphysics which we discussed earlier. Secondly, "πάντα χωρεῖ, οὐδὲν μένει" (all flows, all changes) and should we really be surprised by the changes taking place in the position of even the most important scientic trends in the life of human society? Before our own eyes this has taken place and takes place in biology, especially molecular biology. In general, biology now occupies the 'first place' which belonged to microphysics in the first half of the century. Once again we meet here with 'interference' — the astonishingly interesting and by themselves important, that is, important from a purely scientific point of view, achievements of genetics are at the same time extremely important for medicine, agriculture, and so on.

It seems to me that I am only kicking in an open door, but this was clearly not the case to judge from the criticism which I have already mentioned. It reduces, in the first place, to the problem: How can one be sure that microphysics will not in the future lead to new and important achievements which will have practical applications such as the mastering of atomic (nuclear) energy? After all, 50 or even 45 years ago nobody thought that nuclear physics would ever have any practical applications in the foreseeable future. Of course, it is completely impossible to state that microphysics will not once again climb on the 'world stage' (in a sense which should be clear from the foregoing). However, this is far from certain and, more importantly, this does not change the validity of our estimate of the place of microphysics at the present time.

I feel that the main source of criticism and irritation which I have met in connection with my remarks about the change in position of microphysics is connected with a far from accurate idea about the role and place of microphysics inside physics itself. My third remark will deal with that point.

3. There is absolutely no doubt about the special position of microphysics or, to be more precise, of a number of problems which belong to it, such as problems about the number and properties of quarks and gluons, about grand

unification, proton decay, the neutrino mass, and so on and so forth. One can say the same in astronomy about the cosmological problem and about black holes. It is just that kind of problems which, not without reason, attracts most strongly an appreciable fraction of the most talented young people. The novelty and the fact that these problems are still unsolved attracts them. If we know the foundations — the basic ideas and the appropriate theory — this gives us a certain tranquillity and confidence. However, if we do not know them, we are in a completely different ballpark. To a large extent one proceeds by trial and error. There are in that case no strict rules for the game, gambling reigns, and there is great excitement. What should, however, be considered to be a crude and fundamental error — and one sometimes encounters it — is to treat the rest of physics and astronomy which is not at the front, as of minor, secondary importance.

Take, for instance, solid state physics. The foundations are here well known, one need not think about quarks and one need consider only atomic nuclei, electrons, and light quanta — photons. The fundamental equations which describe all these particles can be assumed to be reliably established in as far as they apply to solid state physics. However, solids are so complicated and consist of so many particles that, as I mentioned already, it took 46 years after the discovery of superconductivity to explain the nature of this phenomenon. And how much effort had not been expended on trying to solve the problem! One can say the same about the problem of phase transitions and about a few others. L.D. Landau has told me that he had not thought as much about any problem as on that of phase transitions, but he was unable to solve it outside the mean-field approximation.

One should also not forget that physics has not fallen apart into separate sections. No, there are many general methods in physics and their number increases all the time, and there appears a lot which is of interest to very many people. Briefly, I think that if we want at all to introduce a class list' in physics one should consider as a minimum three categories of problems. Firstly, those problems in microphysics which are connected with the broadening and deepening of the foundations of physics themselves. Secondly, and this is a rather large number of problems, the very difficult and fundamental problems from all different fields of physics for which one needs not worry about the foundations, but for which it is far from clear how one can solve the problem, what will be the answer, what the results, and so on. Thirdly, we have a class of already more special problems which are, however, sometimes very important for technology and applications. In this case, of course, it may also turn out to be extremely difficult to make progress, but the kind of problems and the style of research are, clearly, closer to technology or to engineering. All this together makes the 'physics' building. It would be absurd to state that only the foundations are important, just as much as it would be, if we were talking about any building, say, a house or the Lindemann Laboratory in which these lectures are given.

To the same extent there is no basis for considering the builders of the foundations to be some kind of high priest, some special caste or élite, while all the other builders of the building are second-rate specialists. I shall not develop these considerations which are sufficiently obvious; they are clearly evoked by the friction and misunderstandings which are sometimes met with among physicists and astronomers of different specialities.

IV. ASTROPHYSICS

Astronomy has always been closely connected with physics. However, sometimes the connection has become closer and sometimes they have drifted apart. The birth of classical mechanics, in general completed by the work of Newton, was to a large extent based upon the astronomical results of Copernicus, Galileo, and Kepler. The use of telescopes in astronomy, started by Galileo, was based on the achievements in optics. Later on, though, the split into specialities – physics and astronomy – became more pronounced and deeper. This was still true 40 years ago and was reflected in many ways. For instance, at Moscow University at that time astronomy belonged to the mechanics-mathematics faculty, while now it is concentrated in the physics faculty. Of course, this example is not particularly impressive. It is more important that nowadays it is difficult to tell of many people whether they are physicists or astronomers, while astrophysics is occupying a dominating position in astronomy. This is connected with the ongoing process – which is already almost completed – of changing astronomy from an optical to an all-wavelength science. The mastering of new ranges – radio-waves, submillimeter waves, X-rays and gamma-rays, led to the drafting of new people, mainly physicists and radio-engineers. On the other hand, the brilliant results and astronomical discoveries required for their understanding a wide use of theoretical physics and in general attracted to them the close attention of physicists – I myself belong to those physicists both by training and through my main field of activity which at one time started to involve astrophysics; my first astronomical paper which was devoted to the radio-emission of the Sun appeared in 1946. Moreover, nowadays it is difficult to draw any distinct boundary between physics and astrophysics – the distinction between them is rather arbitrary. For instance, the general theory of relativity (GR) is first of all physics. However, almost all what is now connected with GR plays, so to speak against a cosmic (astronomical) background. For that reason I have assigned the problems connected with GR to the 'astrophysics' section of our list of key problems. The appearance of such a section was, however, organic and inevitable - otherwise a list of key problems would be clearly one-sided. I feel that, on the other hand, the elimination of problems from biophysics and, say, geophysics, is admissible.

We shall now discuss some of the key astrophysical problems.

17. *Experimental verification and limits of applicability of the general theory of relativity.** The verification of fundamental physical theories and their corresponding predictions nowadays occurs, as a rule, rather fast. A striking example is non-relativistic quantum mechanics. Even if there are still discussions about its completeness — I have in mind the problem of whether it is possible to get rid of probability statements — there are practically no doubts at all about the validity of the results of quantum mechanics themselves. The general theory of relativity (GR) is in this respect a well known exception; it is a decade older than quantum mechanics, but its verification remains a topical physical problem which we included in our list.

The cause for this is very clear. GR is a quite well defined theory of the gravitational field which was constructed by Einstein on the basis of the equivalence principle and which describes the gravitational field using solely the metric tensor g_{ik}, which determines the line element, $ds^2 = g_{ik}\, dx^i\, dx^k$ ($i,k = 0, 1, 2, 3$; $x^0 = ct$ is the time coordinate, and we assume summation over repeated indices). In an inertial (Galilean) frame of reference (if it exists) we have, when we use Cartesian coordinates

$$g_{00} = 1\ , \quad g_{11} = g_{22} = g_{33} = -1\ , \quad g_{ik} = 0 \quad \text{when} \quad i \neq k\ . \tag{9}$$

In a weak gravitational field, which means that the components g_{ik} may be assumed to be close to the Galilean values (9), we can put

$$g_{00} = 1 + \frac{2\phi}{c^2}\ , \quad \nabla^2\phi = 4\pi G\mu\ , \tag{10}$$

where ϕ is the Newtonian gravitational potential; we assume here that 'at infinity' $\phi \to 0$; μ is the mass density. As indicated, the condition for the gravitational field to be weak is the following one:

---

* **For details see V.L. Ginzburg, Sov. Phys. Uspekhi, 22 (1979) 514; C.W. Will, in 'General Relativity' (Eds. S.W. Hawking and W. Israel) Cambridge University Press, 1979.**

$$\frac{|\phi|}{c^2} \ll 1 . \qquad (11)$$

At the Sun's surface

$$\frac{|\phi_\odot|}{c^2} = \frac{GM_\odot}{c^2 r_\odot} = \frac{r_{g\odot}}{2r_\odot} = 2.12 \times 10^{-6} \quad , \quad r_{g\odot} = \frac{2GM_\odot}{c^2} = 2.94 \times 10^5 \text{ cm} , \qquad (12)$$

as the solar mass $M_\odot = 2 \times 10^{33}\ g$ and the radius of the photosphere

$$r_\odot = 6.96 \times 10^{10} \text{ cm} ;$$

the gravitational constant

$$G = 6.67 \times 10^{-8} \text{ g}^{-1} \text{ cm}^3 \text{ s}^{-2} .$$

The quantity r_g introduced in (12) is called the gravitational radius,

$$r_g = \frac{2GM}{c^2} \approx 3 \times 10^5 \frac{M}{M_\odot} \text{ cm} . \qquad (13)$$

For the Earth

$$\frac{|\phi_♁|}{c^2} = \frac{GM_♁}{c^2 r_♁} = 7 \times 10^{-10} \quad , \quad M_♁ = 5.98 \times 10^{27} \text{ g} ,$$
$$r_♁ = 6.37 \times 10^8 \text{ cm} , \quad r_{g♁} = 0.86 \text{ cm} . \qquad (14)$$

For a circular planetary orbit $|\phi|/c^2 = v^2/c^2$ where v is the velocity of planet: for the Earth's orbit $|\phi|/c^2 \sim 10^{-8}$ as $v = 3 \times 10^6$ cm/s.

The gravitational field within the boundaries of the solar system can thus not only be called weak, but even very weak; the first correction terms to the Newtonian theory of universal gravitation are not larger than $2|\phi_\odot|/c^2 \sim 4 \times 10^{-6}$ (see (10) and (12)). To be precise, one of the effects predicted by GR, the deflection of light or radio-wave rays which pass close to the Sun is given by the angle*

---

* Strictly speaking, this effect was predicted already in 1801 (!) by Soldner on the basis of Newton's theory and corpuscular ideas about light. However, in this way one gets a deflection which is only half that predicted by (15). This deflection, smaller by a factor 2, was predicted by Einstein in 1911 on the basis of the equivalence principle, before he had completed his GR. The fact is that the result (15) is obtained (Einstein, 1915) only when we take into account not only the change in the component g_{00} (see (10)), but also the component $g_{rr} = -1 + 2\phi/c^2$ (see (17) below). Physically this means that the deflection of the ray is the consequence not only of the equivalence principle, which leads to $g_{00} = 1 + 2\phi/c^2$, but also of the curvature of space.

It is a curious fact that the first attempt to measure the light deflection effect was undertaken in 1914 by a German expedition going to Russia just before the start of the first world war. Because of the outbreak of war no observations were made, but if they had occurred and turned out to be successful, they would have led to the result (15) and thereby have disproved Einstein's prediction at that time.

$$\alpha = \frac{4GM_{\odot}}{c^2 R} = \frac{2r_{g\odot}}{R} = 1.75'' \frac{r_{\odot}}{R} , \tag{15}$$

where R is the impact parameter or, in practice, the closest distance between the ray and the centre of the Sun (note that a tall person is seen from a distance of around 200 km at an angle of about $2''$).

Two other, sometimes called critical, GR effects indicated by Einstein are the gravitational frequency shift and the rotation of the perihelium of the planets which is largest for Mercury; they are also of order $|\phi|/c^2$. The same is true for the relativistic retardation of signals for the radio-location of planets — this GR effect is genetically connected with the light deflection in the Sun's field — to be more precise, light deflection and signal delay are, so to speak, two sides of the same coin.

For a long time, even already decades, the problem of the experimental verification of GR was reduced to the observation of the above-mentioned effects, and also to a verification with maximum accuracy of the equality of inertial and heavy mass. One may note that all predictions of GR were confirmed and there are no indications at all that there might exist any contradictions of GR. However, the accuracy is not particularly impressive — we are talking about agreement of the theory with observation within the limits of accuracy which reached one percent or, sometimes 0.1 to 0.01%. There are now a number of projects aimed at improving the accuracy of these data. Of course, I do not want to object to this, but as far as I am concerned, I feel that GR in the weak-field domain has been sufficiently reliably verified. This, however, does not guarantee the validity of GR also for strong fields and it is just the verification of GR in strong fields which is a topical, key problem.

To make clear what we are dealing with here, I shall write down the GR metric for a spherically symmetric, non-rotating mass M (Schwarzschild solution, 1916):

$$ds^2 = \left(1 - \frac{r_g}{r}\right) c^2\, dt^2 - \frac{dr^2}{1 - (r_g/r)} - r^2 (\sin^2\theta\, d\phi^2 + d\theta^2). \tag{16}$$

Here $r_g = 2GM/c^2$ (see (13)) and we have used 'spherical' spatial coordinates r, θ, and ϕ — in these coordinates the length of a circle with its centre at the centre of the mass is equal to $2\pi r$. In the weak-field approximation (11), which is in this case equivalent to $r_g/r \ll 1$, the metric (16) has the form

$$ds^2 = \left(1 - \frac{r_g}{r}\right) c^2\, dt^2 - \left(1 + \frac{r_g}{r}\right) dr^2 - r^2(\sin^2\theta\, d\phi^2 + d\theta^2) . \tag{17}$$

This is also the metric in a weak field which we have already mentioned, in which

$$g_{00} = 1 + \frac{2\phi}{c^2} \quad , \quad g_{rr} = -1 + \frac{2\phi}{c^2} \; .$$

It is clear from (16) and (17) that when we look at the Schwarzschild solution the transition from a verification of the validity of GR in a weak field to the verification of its validity in any field is far from obvious. Indeed, according to (16)

$$-g_{rr} = \left[1 - \frac{r_g}{r}\right]^{-1} = 1 + \frac{r_g}{r} + \frac{r_g^2}{r^2} + \frac{r_g^3}{r^3} + \ldots \; , \qquad (18)$$

and it is immediately clear that a knowledge of the first two terms of the series — which corresponds to the weak-field approximation — in no way guarantees the validity of the equation $-g_{rr} = [1-(r_g/r)]^{-1}$. Those theories of the gravitational field which attempt to compete with GR are constructed in such a way that in the weak-field approximation they agree with GR and, especially, lead to expression (17).

To verify GR in a strong field is a very complicated problem. Within the confines of the solar system terms of order ϕ^2/c^4 are already at least 5 to 6 orders of magnitude smaller than the terms of order $|\phi|/c^2$. For neutron stars at their surface $|\phi|/c^2 \sim 0.1$ to 0.3 — for instance, for a mass $M = M_{\odot}$ and a radius $r_n = 10\,\text{km}$, we have $|\phi_n|/c^2 = GM_{\odot}/c^2 r_n \approx 0.15$ — and there are some possibilities of verifying GR in a not too weak field. The same is true for the binary pulsar. However, if we forget for a moment all about cosmology, the problem of the verification of GR in a strong field will be closely connected with the problem of black holes.

Close to the event horizon, which in the coordinates of (16) corresponds to a value $r \to r_g$, the gravitational field is, of couse, a strong one. The possibility of the existence of black holes follows, unconditionally, from GR. Therefore even the discovery of black holes in itself would be an important confirmation of GR and subsequently various quantitative observations might be looked for to verify the validity also of the metric (16) itself, or the more general Kerr metric which refers to stationary rotating black holes. At the same time, as is usually the case, the reverse statement is incorrect: if no black holes are discovered, this would by no means disprove GR. The fact is that the formation of black holes when stars or gas clouds collapse is a very complicated process. The collapse can be prevented by the impossibility to get rid of the angular momentum, by possible fragmentation, or by nuclear explosions. Therefore during a period of around 10,000 million years or even less, if we talk about the age of stars, in the observed expanding Universe the formation of black holes may turn out to be a rather improbable process.

Somehow or other, the problem of discovering black holes is one of the most important and absorbing ones. I feel that this problem or, to be more precise, all problems dealing with black holes, is the second most important one in contemporary astronomy — one may with good reasons consider the cosmological problem to be the most important one. This is the reason why I resolved not to touch upon cosmology and black-hole physics just as I did not touch upon microphysics in the present series of lectures, because of the restricted amount of time and their general nature. Proceeding differently would mean, indeed, to denigrate, belittle these problems which, in truth, are among the most profound and largest problems of our time.*

As far as the theory of gravitation is concerned we limit ourselves in what follows to only a few remarks.

For stars with a radius which is close to the gravitational radius the average density is given by

$$\rho \sim \frac{M}{\frac{4\pi}{3} r_g^3} = \frac{3c^6}{32\pi G^3 M^2} \sim 2 \times 10^{16} \frac{M_\odot^2}{M^2} \text{ g cm}^{-3} . \tag{19}$$

Already for $M \sim 10\, M_\odot \approx 2 \times 10^{34}$ g the density (19) is of the order of the nuclear density $\rho_n \approx 3 \times 10^{14}$ g cm^{-3} for which the equation of state of matter is on the whole quite well known. Besides, for densities somewhat larger than the nuclear density there are also no reasons to expect such radical changes in the equation of state that it could prevent collapse. In that connection limitations of a microphysical nature do not play a rôle for macroscopic black holes (it will be clear from what follows that this means for $M \gtrsim 10^{26}$ g $\sim 10^{-2} M_\oplus$) and the problem of the applicability of GR in a strong field has no connection with quantum effects or the inapplicability of the existing space-time ideas 'on a small scale'.

Indeed, let us assume that the applicability of GR is limited by quantum effects — the occurrence of quantum fluctuations in the metric g_{ik}. It is well known that then GR is not applicable only for length and time scales l and t or for densities ρ which are comparable with the Planck values

$$l_g = \sqrt{(G\hbar/c^3)} = 1.6 \times 10^{-33} \text{ cm} , \quad t_g = l_g/c = 5 \times 10^{-44} \text{ s} ,$$
$$\rho_g = c^5/\hbar G = \hbar/c\, l_g^4 \approx 5 \times 10^{93} \text{ g cm}^{-3} , \tag{20}$$

where $\hbar = 1.05 \times 10^{-27}$ erg s. The density ρ_g and the length l_g correspond to a mass

---

* In the books and papers which I have quoted earlier one can find some information about these problems and references to literature. A whole range of other books and review papers are also easily available about these topics.

$$M_g \sim \rho_g l_g^3 = \hbar / c l_g = \sqrt{(\hbar c/G)} \approx 2 \times 10^{-5} \text{ g},$$
$$M_g c^2 \sim 10^{16} \text{ erg} \sim 10^{28} \text{ eV}. \quad (21)$$

Thus when $l \gg l_g$, $t \gg t_g$, $\rho \ll \rho_g$, $M \gg M_g$, quantum effects do not prevent the application of GR — a completely well defined classical theory of the gravitational field.* If, however, $l \lesssim l_g$, $t \lesssim t_g$, $\rho \gtrsim \rho_g$, and so on, quantum effects will, in general, be large and GR is inapplicable. The creation of a quantum theory of gravitation — which is usually attempted using GR (and not another classical theory of gravitation) as a basis — is is problem of enormous importance and difficulty. The complete solution has not yet been found. It is clear from what we have said that a quantum theory of gravitation is necessary for studying the region near the classical singularity, where $\rho \to \infty$, which occurs in cosmological solutions and where gravitational collapse takes place. It is very probable that singularities need not and will not arise in the quantum theory of gravitation.

In the framework of verifying the validity of GR in a strong field one should not connect this problem with quantum effects. Near the gravitational radius (the event horizon) of a massive black hole quantum corrections are negligibly small and, if GR is valid, one can make well defined predictions — for instance, the metric (16) should be valid for a non-rotating, stationary black hole. Apart from some comparison of theory and observation for strong fields a purely theoretical analysis of the theories of gravitation which try to compete with GR also plays a large rôle. Such theories exist, but usually meet with profound difficulties. For instance, in them — or, at any rate in some of them — there are solutions with a negative energy which leads, in particular, to the following result: The orbit of a binary does not contract as a result of gravitation (that is, the system loses energy), but, contrariwise, it 'uncoils' — the system gains energy. It is clear that such theories are inapplicable and can be scrapped. Strictly speaking, I know of only one theory of the gravitational field which, one might say, generalizes GR and which includes it as a particular or, rather, a limiting case. I am referring to a theory which introduces apart from the tensor g_{ik} also a scalar field. However, the problem of the analysis of 'viable' theories of the gravitational field is very complex and I have not specially occupied

---

* **What we have said here does, of course, not mean that we can neglect quantum effects completely. For instance, black-body radiation (evaporation) of black holes — one of the most important discoveries in physics in the last decade (Hawking, 1974) — may be important also when $M \gg M_g$, but the gravitational field of the black hole can in that case be taken to be the classical field.**

myself with it (for details I refer to the earlier cited paper by C.W. Will).

Let me note in connection with verifications of GR yet the following point. The profoundness, beauty, and simplicity – where we use the term simplicity in some higher sense of the word – of GR, its confirmation in a weak field, its analysis during many years – all this makes it for many people to be of little interest and with little actuality to further verify GR in the classical region, that is, apart from the vicinity of 'true' singularities (no such 'true' singularity exists at the event horizon). I myself belong and have always belonged to the band of fervent admirers (and I am not afraid to use this word) of GR. On the level of intuition and faith, I feel that GR in the classical region is completely valid, that is, its limits are determined by quantum effects or by the possibility – to be discussed later on – of the existence of a fundamental microscopic length l_f which is larger than the gravitational length $l_g \sim 10^{-33}$ cm. However, physics is not based, and cannot be based, only upon intuition and faith; the highest courts are always the experiments and observations. Also, in a purely theoretical framework some possible generalizations of GR – for instance, connected with the introduction of a new scalar field or the related assumption that the gravitation constant G varies with time – do not look too extravagant. Somehow or other, I feel that the verification of GR in strong fields, the problem of the existence of black holes, and the problem of the limits of applicability of GR are all key problems – or, at any rate, one general key problem – of contemporary physics and astrophysics.

Finally, in this connection one last remark. The quantum limitations to the region of applicability of GR which are characterized by the Planck parameters (20) are obtained assuming that the existing space-time ideas are, in general, adequate up to scales $l \sim l_g$ and $t \sim t_g$. However, experimentally, based upon high-energy physics data and data from quantum electrodynamics one can state that "all is in order" only for $l \geqslant 10^{-16}$ cm. Physics has as yet not penetrated into the region of smaller scales. There exists a 'gap' of as much as 17 orders of magnitude between $l \sim 10^{-16}$ cm and $l_g \sim 10^{-33}$ cm. It is therefore completely impossible to deny a meaning to the basis of a problem which has been discussed already for several decades: Does there exist a fundamental length l_f such that for $l \leqslant l_f$ space is no longer 'normal' but becomes granular, quantized, or so on?

The idea that there exists a fundamental length and, especially, a length $l_f \sim 10^{-17}$ cm was used for many years with much support in connection with the occurrence of divergent expressions in quantum field theory. If a fundamental length existed one could think that all 'divergences' should be 'cut-off' at distances (or wavelengths) of order l_f. Such considerations led, when applied to weak interactions, at once to a value $l_f \sim 10^{-17}$ cm (the energy $E_f \sim \hbar c/l_f \sim 10^{12}$ eV). However, these arguments are now no longer valid after the creation of a unified theory of electro-weak interactions. As a result the possibility of the existence of a fundamental length is now rarely mentioned and one boldly extrapolates the theory down to lengths $l \sim l_g \sim 10^{-33}$ cm. Well, boldness in physics sometimes has turned out be to fully justified. On the other hand, the problem of the existence of a fundamental length does not go away, of course, and one should keep this possibility in mind. If there exists some fundamental length $l_f > l_g$, it is the scale length l_f — and not l_g — which, clearly, decides the limits of the region of applicability of GR and of all of the existing theory. From dimensionality considerations it follows that the length l_f corresponds to a density

$$\rho_f \sim \frac{\hbar}{c l_f^4} \sim 5 \times 10^{93} \frac{l_g^4}{l_f^4} \text{ g cm}^{-3} . \tag{21}$$

If $l_f \sim 10^{-17}$ cm, we have $\rho_f \sim 10^{30}$ g cm^{-3} (in (19), if $\rho \sim \rho_f \sim 10^{30}$ g cm^{-3}, we have $M \sim 10^{26}$ g). We note that even, for example, for $l_f \sim 10^{-20}$ cm, $\rho_f \sim 10^{42}$ g cm^{-3} which is 32 orders of magnitude less than the density ρ_g. From this it is clear that the possibility that there exists a length $l_f > l_g$ is far from trivial. I feel that one should not forget this possibility; after all, we know practically nothing about length scales $l < 10^{-16}$ cm from experiments.

Even from what we have said, for all its brevity and fragmentary nature, it is clear that the problem of the singularity, which is first of all, connected with the cosmological problem, turns out to be closely connected with microphysics — in fact, this is a remarkable demonstration of the deep unity of and profound connections between physics and astronomy. Moreover, a study of the 'early Universe', that is, of the earliest phases (we are talking here of time intervals which are removed from the classical singularity only by a time of the order of $t_g \sim 10^{-43}$ s or by a time only a few orders of magnitude longer) of the evolution in cosmological models with a singularity may perhaps be the only way somehow to connect some microphysical theories with observations.*

I have already mentioned that in these lectures I shall not discuss this important problem, or cosmology in general, or black-hole physics —

* See, for instance, the most recent survey I know about this topic: E.W. Kolb and M.S. Turner, Nature, **294** (1981) 521,

although I shall mention black holes in passing in what follows. In the remainder of this section I shall touch only upon problems 18, 21, and 23 from my list. The origin of cosmic rays (problem 22) is the subject of a separate lecture.

18. *Gravitational waves.* Einstein completed the formulation of his general theory of relativity in 1915 — I have in mind both the construction of the equation for the gravitational field, the metric tensor g_{ik}, itself, as well as the explanation of the rotation of Mercury's perihelium and the deflection of light rays passing close to the Sun. Already in the next year, 1916, Einstein studied the problem of gravitational waves and in 1918 he obtained the now well known formula which connects the energy which is emitted in the form of gravitational waves with the square of the third time derivative of the quadrupole moment of the mass of a non-relativistic emitting system, such as a binary.*

The existence of gravitational waves is particularly clear if we bear in mind that the general theory of relativity is a generalization of the Newtonian theory of universal gravitation, analogous to the change from electrostatics to electrodynamics. The speed of gravitational waves, as of electromagnetic waves, is equal to the speed of light in vacuo. Solid bodies deform under the influence of gravitational waves and start to oscillate,

---

* I refer here to the formula

$$-\frac{dE}{dt} = \frac{G}{45c^5} \dddot{D}_{\alpha\beta} \dddot{D}_{\alpha\beta},$$

where E is the energy of the system,

$$G = 6.67 \times 10^{-8}\ \mathrm{g^{-1}\ cm^{-3}\ s^{-2}}$$

is the gravitational constant, and

$$D_{\alpha\beta} = \int \mu\,(3x_\alpha x_\beta - r^2 \delta_{\alpha\beta})\,dV$$

is the mass quadrupole moment tensor; the total mass of the system is

$$M = \int \mu\,dV\,, \quad r^2 = x_\alpha x_\alpha \equiv x^2 + y^2 + z^2\,.$$

The luminosity (power) of the gravitational radiation of a mass distribution (double star, rotating bar or dumbbell, ...) with a quadrupole moment $D \sim Mr^2$, where r is a characteristic distance, such as the radius of the orbit or the length of the bar, is of the order

$$L_\mathrm{g} = -\frac{dE}{dt} \sim \frac{GM^2\omega^6 r^4}{c^5}, \qquad (22)$$

where $\omega = 2\pi/T$ is the frequency and T the period of rotation.

and a system of free particles — for instance, two or several satellites — change their relative distances.

It is well known that the gravitational interaction is the weakest of the known interactions. Here on Earth its effect is strong only because gravitational forces are long range and the whole of the Earth acts upon us. However, if we take two protons as an example, their gravitational attraction is $e^2/Gm_p^2 \sim 10^{36}$ times weaker than the Coulomb repulsion ($e = 4.8 \times 10^{-10}$ cgsu, $m_p = 1.67 \times 10^{-24}$ g). Therefore the luminosity (power) L_g of the gravitational radiation is relatively very small. For instance, for a binary, with stellar masses $M_1 \sim M_2 \sim M_\odot \sim 2 \times 10^{33}$ g, orbital radius $r \sim r_{\odot♁} = 1.5 \times 10^{13}$ cm $=$ 1 au, and square of the rotational frequency $\omega^2 = G(M_1 + M_2)/r^3$, we have

$$L_g \sim \frac{GM_\odot^2\,\omega^6\, r_{\odot♁}^4}{c^5} \sim \frac{G^4 M_\odot^5}{c^5\, r_{\odot♁}^5} \sim 10^{19}\ \text{erg s}^{-1}\ , \tag{23}$$

and even for $r = 2r_\odot = 1.4 \times 10^{11}$ cm the luminosity $L_g \sim 10^{29}$ erg/s, whereas the total solar luminosity is $L_\odot = 3.86 \times 10^{33}$ erg/s.

The flux of gravitational waves from a cosmic object which is at a distance R from us is of the order of magnitude

$$F_g \sim \frac{L_g}{4\pi R^2}\ .$$

Apart from the flux F_g one can use the quantity*

$$h = \frac{1}{\omega_g}\sqrt{\frac{8\pi G\tilde{F}_g}{c^3\,\omega_g}} \sim 2 \times 10^{-19}\sqrt{\frac{F_g}{\omega_g^2}}\ , \tag{24}$$

which in the case of gravitational wave pulses is a more convenient quantity. Here $\tilde{F}_g = \int F_g\,dt$ is the energy of the gravitational wave pulse, per unit area, τ_g the length of this pulse, so that $\tilde{F}_g/\tau_g \sim F_g$, and $\omega_g \sim 2\pi/\tau_g$ is the average frequency. The quantity h is the amplitude of the oscillations Δl produced by the pulse on two test masses which are at a distance l apart; thus $h = \Delta l/l$ while by assumption $l \ll c\tau_g$. For weak gravitational waves on a background of plane space, in which we are interested, the metric has the form

$$g_{ik} = g_{ik}^{(0)} + h_{ik}\ ,\quad g_{00}^{(0)} = 0\ ,\quad g_{\alpha\beta}^{(0)} = -\delta_{\alpha\beta}\ ,\quad g_{\alpha 0}^{(0)} = 0\ ,\quad |h_{ik}| \ll 1\ ,$$

---

* See, for instance, D.H. Douglass and V.B. Braginsky, in "General Relativity (An Einstein Centenary Survey)" (Eds S.W. Hawking and W. Israel), Cambridge University Press, 1979; V.B. Braginsky and V.N. Rudenko, Phys. Reports, **46** (1978) 165.

and h is the amplitude of the waves h_{ik}; we do not discuss here the problem of the polarization of the waves and factors of order $\frac{1}{2}$.

For a binary with $L_g \sim 10^{29}$ erg/s, $R \sim 10$ pc $= 3 \times 10^{19}$ cm, and $\omega_g \sim 3 \times 10^{-4}$ s^{-1} we have $F_g \sim 10^{-11}$ erg/cm^2 s and $h \sim 2 \times 10^{-21}$. For the pulsar PSR 0532 in the Crab Nebula the total power of emission in all bands of electromagnetic waves $L \sim 10^{38}$ erg/s and the luminosity L_g cannot exceed this quantity, as otherwise the rotation of the pulsar would be slowed down faster than is observed. In fact, one can take it that $L_g \ll 10^{38}$ erg/s, but even if we put $L_g \sim 10^{38}$ erg/s, we have $F_g \sim 3 \times 10^{-7}$ erg/cm^{2}s, as the corresponding distance $R \sim 2000$ pc $= 6 \times 10^{21}$ cm (in this case $\omega \sim 200$ s^{-1} and $h \sim 5 \times 10^{-25}$).

As far as we know the measurement of the above-mentioned small fluxes is at the moment not yet a realistic possibility. The situation is better with respect to pulsed sources. When a star collapses, or when we have a supernova outburst, or a similar event, we may expect energy emission in the form of gravitational waves which may reach $W_g \sim 10^{52}$ to 10^{54} erg — we bear in mind that $M_\odot c^2 \sim 10^{54}$ erg — with a pulse length $\tau_g \sim 10^{-2}$ to 10^{-4} s, that is, with $\omega_g \sim 2\pi/\tau_g \sim 10^3$ to 10^5 s^{-1}. To be precise, when $W_g \sim 10^{52}$ erg and $R \sim 3$ Mpc $= 10^{25}$ cm (within the corresponding volume there are nearly 300 galaxies and one may expect 1 to 10 supernova outbursts per year) $F_g \sim 10^{52}/4\pi R^2 \sim 10$ erg/cm^2, $F_g \sim 10^3$ to 10^5 erg/cm^{2}s and $h \sim 10^{-20}$ to 10^{-21}. In general, from the events which we have considered it is diffivult to expect bursts with $h > 10^{-19}$, when $\tau_g \sim 10^{-3}$ s.

The existing gravitational antennae of the first generation cannot receive such pulses — their sensitivity is insufficient. The second generation antennae which are being constructed — in particular, from sapphire — are designed to measure displacements $\Delta l \sim 10^{-17}$ cm, which is a possibility which can already be attained. Hence, for an antenna length $l \sim 100$ cm we have already $h = \Delta l / l \sim 10^{-19}$. Therefore, if we are lucky, that is, if not less than one event per year with $h \geqslant 10^{-19}$ reaches the Earth, gravitational waves may successfully be detected in the very near future. It is very well possible that for actual bursts $h \sim 10^{-20}$ to 10^{-21} and then one must wait still a little bit longer before we can detect them, but probably this will still occur in the present century.

The detection of gravitational waves of cosmic origin and in that way the birth of gravitational-wave astronomy is first and foremost of astronomical interest. Indeed, in that way yet another channel is opened for astronomical information. Of course, this information does not duplicate at all that obtained through other channels. To be concrete, it is clear from what we have said that a pulse of gravitational waves characterizes changes in the quadrupole moments of stars, of a system of two stars, and so on. At the same time, the problem of gravitational waves is closely connected with the verification of GR. The fact is that in theories of the gravitational field which differs from GR the emission of gravitational waves proceeds, generally speaking, differently from GR. Moreover, these waves may turn out to be no longer purely transverse. At the present time when it is not yet possible to detect gravitational waves it is particularly important that it is possible to verify formula (22) for the power of gravitational radiation, indirectly, so to say. In fact, observations of the changes in the elements of the orbit of the binary pulsar PSR 1913+16 which have been performed during the last six years have given evidence in favour of the validity of formula (22) and thus also of GR.* The facts reduce to the following: as a result of the emission of gravitational waves the binary system naturally loses energy and the stars, one of which is a radio-pulsar, get closer to one another, and the period of their orbital motion decreases — to be precise, for the pulsar PSR 1913+16 the decrease in the orbital period is 7.6×10^{-5}s per year. The analysis of the observational data is in full agreement with GR (equation (22)) and contradicts the predictions of some of the other theories of the gravitational field. It is true, as is usual in such cases, that observations of but a single object are as yet insufficient for a completely confident judgement — one can always assume that the agreement between theory and observations on a single object is fortuitous. Somehow or other, the quoted results for the pulsar PSR 1913+16 make the validity of GR yet more probable, at least, when applied to a range of problems which are connected with gravitational-wave astronomy. Therefore we can have the more confidence to take as our basis the results of GR for planning gravitational-wave detectors, to estimate the power of their emission, and so on.

It is difficult to doubt that the problem of gravitational waves and gravitational-wave astronomy will remain for several years yet among the key problems of astronomy.

21. *Quasars and galactic nuclei. Formation of galaxies.* Pulsars were discovered in 1967, although the first publication appears at the beginning of 1968 and their nature became clear less than two years later. Pulsars are rotating magnetized neutron stars.† Both the occurrence of fast rotation — for

---

* See, for instance, J.M. Weisberg, J.M. Taylor, and L.A. Fowler, Scientific America, **245**, No.4 (1981) 66; see also Ap. J., **253** (1982) 908.

† I am not excluding the possibility that there may exist long-period pulsars which are white dwarfs, but this reservation does not change the situation.

the 'fastest' of the known pulsars, the pulsar PSR 0532 in the Crab, the period $T = 0.033$ s, and of a huge magnetic field ($H \leqslant 10^{13}$ Oe) are completely natural for neutron stars — or, better, are easily explainable, on the basis of very simple considerations about conservation of angular momentum and magnetic flux when an 'ordinary' star becomes a neutron star — we have already discussed this in connection with problem 7. The appropriate conclusions were reached in the literature also before the discovery of pulsars. Without wishing at all to denigrate the rôle of theoretical work and not trying to conceal that my own work in the field of pulsar emission was not particularly successful, I should like to make the following remark. If the short-period pulsars in the Crab and Vela had not been discovered soon after the first pulsars, and it is only those two which definitely had to be connected with the model of rotating neutron stars, the debate about the nature of pulsars would have continued for many years — the white dwarf model would then have competed with the neutron-star model.

This diversion, devoted to pulsars, was made because the fate of quasar studies is completely different. Even if we forget about earlier indications — one might say, ancient history or prehistory — and consider the measurement of the red-shift in the spectrum of the quasar 3C 273, quasars were discovered in 1963, that is, several years before the pulsars. However, so far the nature of quasars or, more precisely, their cores, remains unexplained. This does not mean that during almost twenty years after the discovery of quasars little was done. On the contrary, there has been and is a great deal of activity, but simply we have not found or turned up the necessary key which would have enabled us to open the 'black box', as the core of a quasar or active galactic nucleus is sometimes called.

Before we explain the meaning of this term let us devote a few words to what one may take to be established. Quasars are at cosmological distances — by this we understand that the red shifts of the lines observed in the quasar spectra are of cosmological origin — caused by the fact that the quasars go away from us because they take part in the expansion of the Universe. This problem has provoked arguments, but I — like most astronomers — have never seen any real basis whatever to doubt the cosmological nature of the red shift in quasar spectra, and nowadays such a point of view does not cause any doubts.*

---

* It is something else that it is possible, in principle, that there is also a non-cosmological red shift which in some special cases — for some objects — could even be dominant. Such a reservation does, however, not change the situation with respect to the overwhelming majority of quasars and active galactic nuclei.

Quasars in the original sense of this word (quasi-stellar radio-sources, QSR) turned out to be only a relatively small part of a much broader class of objects (QSO) — the quasi-stellar sources. Moreover, it turned out that quasars and quasi-stellar objects are nuclei of gigantic galaxies, that is, they are surrounded by stars. Similar active nuclei are found in many galaxies — for instance, the well known Seyfert galaxies are gigantic spiral galaxies with active nuclei. For the sake of simplicity we shall in what follows combine all these objects and speak about quasars — compact (radius $r \sim 10^{16}$ to 10^{17} cm) extremely bright sources — whose integrated luminosity reaches 10^{48} erg/s — the luminosity of our Galaxy $L_G \sim 10^{44}$ erg/s.

There is no doubt that the energy output in quasars is connected with the release of gravitational energy. For instance, when a mass M is compressed to a size of the order r — from a size $r_0 \gg r$ — energy of the order

$$\frac{GM^2}{r} \sim Mc^2 \frac{r_g}{r} \sim 10^{54} \frac{M}{M_\odot} \frac{r_g}{r} \text{ erg}, \quad r_g = \frac{2GM}{c^2} = 3 \times 10^5 \frac{M}{M_\odot} \text{ cm}, \tag{25}$$

is released. The total energy release in the case of quasars reaches, apparently, 10^{61} to 10^{62} erg (a luminosity up to 10^{48} erg/s during 3×10^5 to 3×10^6 years). Even for a black hole, when we can put $r \sim r_g$, the mass of the nucleus must reach $10^8 M_\odot$. In general, there is no doubt that for quasars and active nuclei the mass M reaches 10^8 to $10^9 M_\odot$, and may be even somewhat larger (the total mass of our Galaxy $M_G \sim 10^{11} M_\odot$).

Apart from emitting radio-, infrared, and optical radiation, many quasars are also sources of strong X-ray radiation. For instance, in one of the lists of 111 quasars examined by the cosmic X-ray observatory 'Einstein' (the HEAO B ≡ HEAO 2, launched on 13th November, 1978) 35 quasars turned out to be sources of observable X-rays, with photon energies in the range

$$0.5 < E_X < 4.5 \text{ keV}$$

and a luminosity

$$L_X \sim 10^{43} \text{ to } 10^{47} \text{ erg/s}.$$

Apparently, quasars can turn out to be also strong gamma-ray sources although it seems that this has been reliably established only for the quasar 3C 273, with a luminosity $L_\gamma (50 < E_\gamma < 500\,\text{MeV}) \sim 2 \times 10^{46}$ erg/s when $L_X \sim 10^{46}$ erg/s.

What is this radiating nucleus with a size $r \sim 10^{16}$ to 10^{17} cm? The emitting region itself is very probably not in any extra-ordinary condition. However, the 'furnace' (the 'engine', energy source) is hidden inside that region and cannot be seen, while the nature of this 'furnace' to a large extent is unimportant, in the sense that it only weakly affects the characteristics of the observed electromagnetic radiation. The term 'black box' is appropriate in this connection. This is the reason why so far the contents of this box are not clear.

Most often three models for the cores of quasers and active galactic nuclei are discussed:* dense stellar clusters, magnetoids or spinars, that is, a magneto-plasma, rotating 'superstar' without a massive black hole in its centre, and a massive black hole. Of course, it is possible to combine these models. The stellar cluster model is the least likely one, as in this model — at least in its simple variants — separate intensity oscillations in the emission of the quasars are connected with stellar outbursts or, say, with collisions of two stars. However, in that case one can hardly explain the observed large luminosity variations. Moreover, with respect to the stellar cluster model one can make the same objection as for the magnetoid (spinar) model, namely, that it is unclear why such a structure does not ultimately lead to a black hole. For this reason and 'from general considerations' the massive black hole model is nowadays the most popular one.

I have already mentioned that personally I believe in the validity of GR — in the classical region which is sufficient for the present case — and do not doubt that in the framework of GR black holes can exist. Nonetheless the massive black hole quasar model is far from having been proved. Even, if we do not doubt the validity of GR in strong fields there might not be sufficient time — as we also already mentioned — for the formation of a massive black hole. Furthermore, there are no direct indications of the existence of black holes in galactic nuclei — the black box is still sealed. Moreover, if it were easy to form black holes one could expect them to occur also in the nuclei of galaxies which are at the present moment (of course, we are talking about the time when we observe them on Earth) inactive and, in particular, in our own Galaxy. In principle, the presence of such 'dead' (non-active) black holes is not excluded. In fact, their activity is connected with the accretion of gas or of stars which are in the neighbourhood of the hole, and the latter might at some time not be present in sufficient numbers. However, those data which I am aware of about the centre of our Galaxy contradict such an assumption and, apparently, give evidence† against the presence of a black hole with a mass $M \geqslant 10^3 M_{\odot}$ (it is true that the corresponding analysis is

---

* See, for instance, V.L. Ginzburg and L.M. Ozernoy, Astroph. Space Sci., **48** (1977) 401; R.D. Blandford and K.S. Thorne, in "General Relativity' (Eds. S.W. Hawking and W. Israel), Cambridge University Press, 1979.

† V.G. Gurzadyan and L.M. Ozernoy, Astron. Astrophys., **86** (1980) 315; **95** (1981) 39.

still not generally accepted and the question of the existence of a massive black hole in the centre of our Galaxy is still open to discussion). There are objections against the possibility of the existence of massive black holes also in some other galaxies. Moreover, in the framework of the black hole model it is somewhat more difficult than in the magnetoid model to explain the recurrence of activity — the black hole cannot disappear and for the occurrence of a repeated phase of activity one needs very special, albeit not impossible conditions.* However, there are also obscurities in the magnetoid model — the problems of its stability and further fate are insufficiently elucidated. Of course, one must study this problem on the basis of observations and theoretical analysis. At this moment it is important for us only to emphasize that the problem of the model of the cores of quasars and of active galactic nuclei is to a large extent still open and its solution is a topical and very difficult problem.

One of the possible approaches is the observation of the periodicity of the quasar luminosity oscillations. In the magnetoid (spinar) model such oscillations are, in particular, connected with the rotation and they are completely natural and more stable than in the black hole model with an accretion disk. One can, in principle, use also neutrino astronomy to distinguish a black hole from a magnetoid (*vide infra*).

When shall we succeed to explain with confidence the nature of the 'black box', the quasar core? Perhaps it will still take many years for this to happen, but we might also be lucky and, at least in one case (one example), be able to open the black box and find out what it contains.

In the title of the present section (problem 21 of our list) the question of the formation of galaxies also appeared. In some sense it is connected with the problem of the nature of quasars and galactic nuclei, but on the whole they are unconnected. We are here concerned rather with those 'initial' perturbations which develop in the expanding Universe in such a way that in the appropriate stage they lead to the formation of galaxies and of clusters of galaxies. The problem is still far from having been solved. New points are here: firstly, taking into account the contribution from the neutrinos, if the neutrino mass is sufficiently large (say,

$$m_\nu \geqslant 10 \text{ eV} \sim 10^{-32} \text{ g} \sim 10^{-5}\, m \,, \quad \text{where } m = 9 \times 10^{-28} \text{ g}$$

is the electron mass); secondly, a discussion of the possibility that the initial perturbations are purely quantal by nature and are completely caused

---

* See in this connection, for instance, the recently published paper: R.H. Sanders, Nature, **294** (1981) 427.

by the inapplicability of the classical theory of gravitation near the singularities which are part of this theory.

From what I have said one can imagine how interesting and important the problem of the formation of galaxies is; however, a more detailed discussion of it is here not possible.

23. *Neutrino astronomy.* Practically all knowledge about the cosmos, all astronomical information, is obtained by detecting electromagnetic radiation. Moreover, in ancient times and up to the middle of this century it would not have been an exaggeration to replace 'electromagnetic radiation' by 'light' or even 'visible light'. Today's situation is different. There has taken place and there continues to take place a great process of transforming astronomy from optical to all-wavelength astronomy. Radio-astronomy, and in the last decade also X-ray astronomy, have already become somehow the equal of optical astronomy. The sub-millimeter and the even more important and broad region of gamma-ray astronomy only lag behind a little. However, one may think that in the current decade or certainly before the end of the century they will overtake optical astronomy in their development. Of course, the word 'overtake' must here be understood with reservations. The importance of conquering new bands lies in the fact that in each of them one gets specific, new information. There is some overlap, but basically the pictures of the heavens in radio-waves, in optical wavelengths, in X-rays and in gamma-rays are different pictures. This is, of course, well known and obvious. As an example and reminder I mention that the brightest radio-source Cassiopeia A, a supernova remnant, was not at all noted reliably in the optical band, and even now it can be studied optically only with the best telescopes.

Apart from electromagnetic waves we also get — besides meteorites — charged particles from the cosmos — the cosmic rays. Their study is of considerable astronomical and physical interest, and I shall discuss this separately (Chapter 3). From the known remainder there rest the observation of gravitational waves (gravitational-wave astronomy) and the observation of cosmic neutrinos. We shall now make a few remarks about the corresponding field — neutrino astronomy.

There exist already three branches in neutrino astronomy: solar neutrino astronomy — the observation of neutrinos from the Sun, the search for neutrino bursts from supernova outbursts and the like, and neutrino astronomy with high or very high energies.

The Sun and other 'hot' stars obtain their energy basically from nuclear reactions (the rôle played by gravitational compression is usually unimportant and does not last long). Nuclear reactions takes place, of course, in the interior of the stars, near the centre where the temperature is highest. Neutrinos are, however, able to pass through the star and that is a unique property of them (of course, gravitational waves have an even better penetrating ability). The problem of the observation of solar neutrinos not only has been discussed for rather a long time, but already for more than a decade observations have been made (albeit in only one place in the USA). We shall therefore give only a short summary.*

So far attempts to register solar neutrinos have been made only using a chlorine detector — using the reaction

$$^{37}\mathrm{Cl} + \nu_e \longrightarrow {}^{37}\mathrm{Ar} + e^- . \qquad (26)$$

Such a detector registers efficiently only neutrinos with energies $E_\nu \geqslant 0.81$ MeV which are emitted by the Sun in only a relatively small number, mainly coming from the decay of the ^{8}B nucleus. The flux of such neutrinos is very sensitive to the temperature in the solar interior and, in general, to the solar model. Rigorous predictions of the neutrino flux to be expected from the Sun and which can be detected by the reaction (26) are thus difficult to make. The latest theoretical value of the flux obtained from more or less 'standard' solar models is 8 ± 3 SNU (SNU is a solar neutrino unit; when there is a flux of 1 SNU, 10^{36} nuclei of the target nucleus, in this case ^{37}Cl, captures as the result of the reaction, on average one neutrino per second). The latest data from the above-mentioned observations give about 2 SNU (the value 1.95 ± 0.3 SNU is given). This difference of approximately a factor four between calculations and observations has produced an enormous literature. Recently, the discrepancy is rather often connected with hypothetical neutrino oscillations (the reciprocal transformations of electron-, muon-, and, possibly, tau-neutrinos, ν_e, ν_μ and ν_τ into one another). However, I, and many other people, feel that in actual fact there may be some doubt about this discrepancy, as one is dealing with a neutrino flux which is particularly sensitive to the model of the solar interior. On the other hand, almost the whole of the neutrino flux from the Sun which (at the Earth) equals 6.1×10^{10} neutrinos/cm^2 s and which is caused by the reaction

$$p + p \rightarrow d + e^+ + \nu_e , \qquad (27)$$

is practically independent of these models (we note that the neutrino flux

---

* See J.M. Bahcall, Space Sc. Rev., **24** (1979) 227; the latest information I am aware of is contained in preprints which are to be published in the Proceedings of the 1981 Neutrino Conference in Maui, Hawaii (Ed. R.J. Cence).

from the earlier mentioned reaction $^8B \rightarrow {}^8Be + e^+ + \nu_e$ is only of the order of 10^7 neutrinos/cm^2 s).

The neutrinos from the reaction (27) can successfully be registered by a gallium detector (the isotope ^{71}Ga changes into ^{71}Ge with a neutrino threshold energy of only 0.23 MeV; the neutrino energy in the reaction (27) reaches a value of 0.42 MeV). The experiment with gallium requires a detector with a mass of 30 to 50 tons and the gallium is practically retained (such an experiment is planned both in the USSR and in the USA). One can also use other detectors (^7Li, ^{115}In, and others). We shall, unfortunately, still have to wait several years before the new detectors are working. However, taken by and large, solar neutrino astronomy has been born.

The working apparatus for the detection of the reaction (26) for the neutrinos from the Sun, like some others, is able to register the neutrino emission which must occur when stars in our Galaxy collapse (we are dealing with neutrinos from a number of nuclear reactions with $E_\nu \leqslant 10$ MeV). So far such events have not been detected. This indicates that collapses in the galactic disk do not, on average, occur more often than once every few years (this agrees with other estimates). However, in future there is every reason to expect the observation of neutrino bursts from collapsing stars.

Finally a few words about high-energy neutrino astronomy. Neutrinos with high energies, say, larger than 1 GeV are produced practically exclusively by the proton-nuclear component of the cosmic rays. The situation is in that respect similar to the production of π^0-mesons in cosmic rays (such π^0-mesons which decay fast into gamma-photons are important in gamma-ray-astronomy). There exist projects, we have especially the project DUMAND (Deep Underwater Muon And Neutrino Detection) in mind, for detecting high-energy neutrinos under the earth and under water. In the case of DUMAND the registration must occur at a depth of 5 km in the ocean using optical detectors (one will detect the Cherenkov emission of a cascade of particles produced by the neutrino). The energy threshold of the apparatus depends, of course, on its dimensions but is, apparently, inevitably rather high — of the order of 10^2 to 10^3 GeV. The angular resolution reaches 1° — this is a true neutrino telescope. For neutrinos with very high energies, $E_\nu > 10^5$ to 10^6 GeV, deepwater projects with acoustic methods of detection are discussed.

Possible galactic high-energy-neutrino sources ($E_\nu > 10^2$ to 10^3 GeV) are young (up to one year old) supernova shells and 'hidden' sources — pulsars and black holes surrounded by a large thickness of matter. One can, in principle, also detect extra-galactic sources: quasars, active galactic nuclei, galaxies in the stage of their formation (one supposes that this took place for red-shift values $z = (\lambda - \lambda_0)/\lambda_0 \sim 10$ to 30; for the furthest known quasars $z < 4$). We restrict here our remarks to saying also that using neutrino emission and its relation to the gamma-luminosity of quasars and active galactic nuclei one can, in principle, distinguish a black hole from a magnetoid (in the case of a black hole surrounded by a dense shell the neutrino flux can be large while the flux of gamma-rays with $E_\gamma > 70\,\text{MeV}$ is small; for a magnetoid the gamma-ray flux is larger than some minimum value, calculated from the data of the neutrino flux).* Unfortunately, such a method does not 'work' for all models of massive black holes and, more importantly, we shall have to wait at least several years before the large-scale DUMAND project is realized.

I must still mention one more exceptionally important problem in neutrino astronomy — attempts to observe relict neutrinos, which were formed in the relatively early phases of the cosmological evolution (similar to the relict thermal microwave emission, but altogether appreciably earlier). This problem is extremely attractive and important in connection with the assumption that the neutrino mass is not equal to zero, and the neutrinos may play a prominent rôle in cosmology (if $m_\nu \gtrsim 1$ to 10 eV). However, I do not know of any realistic attempts leading to a direct detection of such neutrinos.

One decade turned out to be insufficient to establish neutrino astronomy which should not cause surprise in view of the complexity of the corresponding observations. However, I think that still in this century neutrino astronomy will become an actively operating and broad field of investigations.

V. A FEW REMARKS ABOUT THE NATURE OF SCIENTIFIC REVOLUTIONS IN PHYSICS AND ASTRONOMY

I start with a digression. As far as I can see — and it is true that my observations are very limited — young physicists and astronomers are rarely actively interested in the history of science or its methodology. Apparently this is explained by the fascination of the science itself — that is, of physics and astronomy (I restrict myself only to those sciences as I have only tenuous connections with other fields). Moreover, as I already mentioned, "physics is the game of the young" and even apart from its fascination, young people tend to strike iron while it is hot and they simply have no time for other occupations. I experienced all this myself and although in

---

* See V.S. Berezinsky and V.L. Ginzburg, Monthly Not. Roy. Astron. Sci., **194** (1981) 3.

my student years — already more that forty years ago — I was also interested in philosophy, I did not take it seriously and I remained uneducated. I also did not realize that it is much easier to reach the "leading edge" in physics and astronomy, while in the field of philosophy and history one needs also to possess a high level of general culture and master an enormous amount of factual material, even apart from the fact one must develop a taste for and interest in the analysis of the processes of the development of science and so on.

The advice I received in my time from the well known physicist Leonid Isaakovich Mandel'shtam played for me an important rôle in getting to understand this. His name is little known in the West, and I can only deplore this.* L.I. Mandel'shtam (1879—1944) was an excellent, profound physicist and a charming, but very modest human being. With G.S. Landsberg he discovered, independent of Raman, the combinational scattering of light (in the West this effect is exclusively called the Raman effect, while in the Russian literature one uses both the term Raman effect and combinational scattering). L.I. Mandel'shtam indicated also, independent of Brillouin, the appearance of a doublet in the Rayleigh scattering spectrum and he published a large number of other important papers in optics, radio-physics, radio-technology, and so on. Shortly before his death, when Mandel'shtam already hardly left his house, one of my teachers, I.E. Tamm, who was very close to Mandel'shtam (strictly speaking, he considered Mandel'shtam his teacher and friend) arranged it so that I received an invitation to visit Mandel'shtam. We talked about science, mainly about one problem in plasma physics, but at the end of the conversation I made a remark of a philosophical nature. Mandel'shtam was interested in the philosophy of science; he was a very educated man also in that field. And, apparently, he at once noted that one could not at all say that about me. Somehow or other he smiled and rather mildly, but very definitely he gave me the following advice: "As long as you are young, occupy yourself with concrete physical problems. However, when you are 60 or 65 years old, then is the time for the philosophy, the history of physics, and so on." I believe that he added also that with advancing age it becomes difficult and/or less interesting to solve particular problems.

From my own case I have verified the fact how right Mandel'shtam was. Perhaps I should apologize for this autobiographical remark, but it seemed to me that it may be of interest to my audience.

---

* Unfortunately, I can here refer only to a book in Russian: "Academician L.I. Mandel'shtam; on the occasion of the centenary of his birth", Nauka, Moscow, 1979.

Somehow or other I have occupied myself during the last ten years more and more with general problems of the development and history of science and I have even published some papers on those subjects.* Lecturing here about key problems in physics and astrophysics I felt that it might be appropriate to finish with a methodological kind of remark or, to be more precise, to consider the problem of the nature and contents of the concept of a 'scientific revolution'.

There are various arguments in the literature both about the contents of this concept and also about the concrete estimate of some particular 'revolutionary' events in science. I do not want to state that these arguments are indeed important; when a physicist hears them he feels in those cases that much of it is scholastic or a play with words. However, I somehow was drawn into this discussion in connection with a request to write a review of the Russian translation of Kuhn's well known book.† Moreover, I have for some time felt (and I have written about this) that in astronomy there have been only two scientific revolutions: the first one connected with Galileo's name was caused by the use of telescopes, and the second was when we had the change from optical to all-wavelength astronomy. Such a concept seems to me now to be one-sided and in need of revision.

Kuhn defines in his book a scientific revolution as follows:

> "A revolution is for me a special sort of change involving a certain sort of reconstruction of group commitments. But it need not be a large change, nor need it seem revolutionary to those outside a single community, consisting perhaps of fewer than twenty-five people."

Meeting this position involuntarily reminds me of the 'concept' of a storm in a teacup. Of course, as always when we are dealing with terminology problems, it is impossible to prohibit or to show the untenable of almost any definitions. Between a storm in an ocean and in a teacup there lie an innumerable number of cases differing in scale (storm in a bay, in a lake, in a pond, in a basin, and so on). Revolutions in science can be divided, and this has been done, into global, local, micro-revolutions, and so on and so forth. Ultimately, one can call in fact, any abrupt change or 'reconstruction of group commitments' a scientific revolution. However, such an approach seems to me to be inappropriate. We intuitively connect with the concept of an authentic revolution a drastic change, a large scale upheaval. In this connection

---

* See, for instance, V.L. Ginzburg, Qu. J. Roy. Astron. Soc., **16** (1975) 265; and particularly V.L. Ginzburg, Voprosy filosofii, **12** (1980) 24 (in Russian).

† T.S. Kuhn, "The Structure of Scientific Revolutions", Second Edition, University of Chicago Press, 1970.

it is natural to use the word scientific revolution for a profound change in the natural sciences or in some part of it (such as physics, astronomy, or biology). A further stretching out of revolutions into a procrustean bed of hierarchical stages (global, local, ... , revolutions) hardly produces anything. The real problem which is of some interest is connected with the actual analysis of the real history of and situations in science.

Starting from such a position I shall allow myself to express my opinion about revolutions in physics and astronomy. For instance, in astronomy, if we forget about its establishment in antiquity (we are thinking about the origin of the Hipparchus-Ptolemeian astronomical system) so far only two revolutions have occurred. The first of these happened in the sixteenth and seventeenth centuries and can be connected first and foremost with the names of Copernicus and Galileo. Clearly, we are dealing here with the change from the geocentred to the heliocentred view and the change in observational methods (application of the telescope) developed by Galileo, which led to a whole number of brilliant discoveries. The second astronomical revolution occurred (and, strictly speaking, is still taking place) in our century and is also 'two-headed'. On the one hand, we must here point to the discovery that our Universe is not stationary and the general establishment of extragalactic astronomy. On the other hand, the contents of the second astronomical revolution is connected with the change from optical to all-wavelength astronomy. The latter, as in Galileo's time, has led to a number of remarkable discoveries which we have already mentioned.

In such an approach (or, if you please, classification) the duration of each revolution is somewhat increased but this seems to us to achieve harmony and, more importantly, to be in accordance with the historical truth.

I shall give an analogous classification in physics. If we forget again what happened in antiquity, two revolutions occurred in physics. One of them is connected with the birth of classical mechanics (Copernicus, Galileo, Kepler, and, in particular, Newton) and, strictly speaking, the whole foundations of that building which one usually calls classical physics. Nobody should underestimate the value and enormous achievements of Faraday, Maxwell, and the other creators of electrodynamics, but it seemed in the last century that electrodynamics does not contradict classical mechanics. It is therefore appropriate to connect the second revolution in physics only with the creation of relativity theory and quantum mechanics. The problem about dates is in such cases usually open to discussion. I think that the second revolution in physics most properly should refer to the period 1900 to 1932.

What I have said is rather trivial and I suspect that many have thought so and still think so. However, this opinion differs strongly from Kuhn's position and that is the reason why I thought it appropriate to dwell upon it.

If we stick to the point of view indicated here there arises, naturally, the problem of the character of the contemporary development of physics. Are we living in a period between revolutions or has the third revolution started already? Or — and this is also well possible — has the nature of developments in physics and of the whole of the natural sciences changed in such a way that it is no longer possible or proper to talk about scientific revolutions such as the preceding two? Such problems are not devoid of interest. I made a few remarks in this connection in my 1980 paper which I have quoted earlier and which was published in a philosophical journal. In particular, in that paper, using concrete examples I tried to demonstrate the uselessness of Kuhn's scheme with its "transformations of paradigms", "normal science", "anomalies", "extraordinary investigations", and so on when applied to a whole range of processes of scientific development in the past and in the present. Possible exceptions are just the genuine (in the earlier indicated sense of the word) scientific revolutions on a gigantic scale. However, if we are not dealing with authentic scientific revolutions, but have in view more modest processes in the development of science, it is more appropriate to speak of a struggle between different hypotheses, or about the competition between "research programmes". I am afraid that if I went into details, it would take us too far and it would be inappropriate in the framework of the present lectures.

In conclusion I wish to express my hope that the discussion which I undertook of some key problems in physics and astrophysics was of some interest, especially for students. As far as — physics and astrophysics — colleagues are concerned who are themselves already actively engaged in science, for them these lectures can pretend only to give some illustration of the possibilities which there are for discussing key problems. As I already stated at the beginning, I consider such lectures to be a useful device for widening one's horizon. If somebody does not agree with my exposition with respect to the general approach or with respect to some actual details I wish to urge them to give constructive criticism — giving an interpretation from other positions, from another point of view, and so on. Unfortunately, I have encountered, on the contrary, only non-constructive criticism or, to speak bluntly, resentment. However, if we do not give up our goal itself — to promote the idea of the unity of physics, to gain an understanding of the existence of connections, and to look for ways to develop physics and astrophysics — a constructive approach is the only one which is of use.

CHAPTER 2

The Problem of High-temperature Superconductivity

I. INTRODUCTION AND STATEMENT OF PROBLEM

Superconductivity was discovered in 1911 and this discovery had, one can say, at once a completely distinct character, Indeed, already in the first experiments a steep fall in the resistivity of the sample (mercury) was observed by at least four to five orders (Fig. 2.1). One might think, therefore, that the potential importance of superconductivity for electrical technology became clear without any special delay. Very soon, however, it became clear that the superconductivity of pure metals is normally destroyed already in a very weak critical field H_c (for instance, for mercury even close to the absolute zero $H_c \approx 400$ Gauss). Because of this and for a number of other reasons the study of superconductivity remained for many years a purely physical problem which was not connected with technical applications. The discovery of superconducting alloys with huge critical fields and critical currents, the progress in the field of cryogenic technique, and also

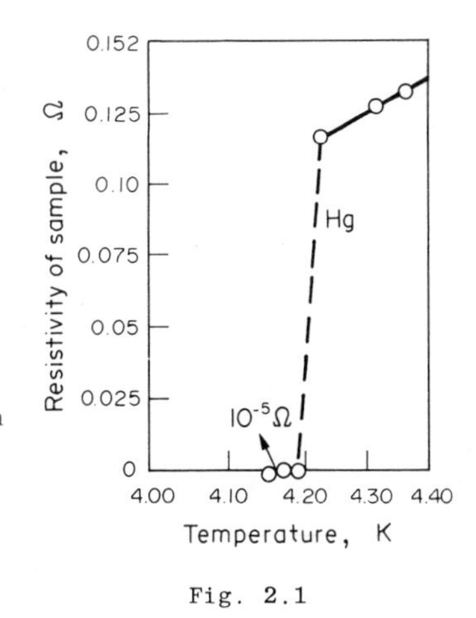

Fig. 2.1

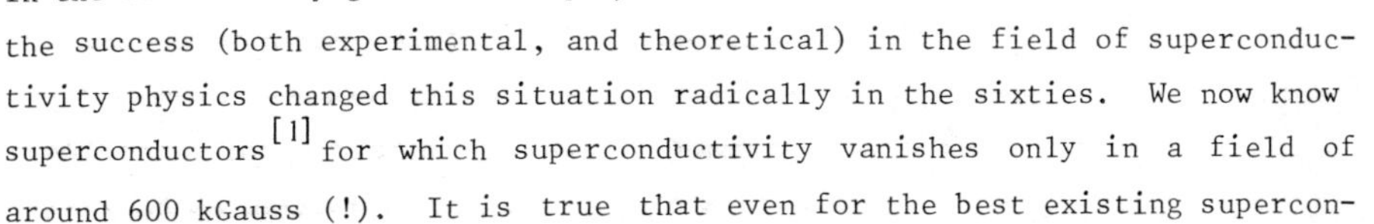

the success (both experimental, and theoretical) in the field of superconductivity physics changed this situation radically in the sixties. We now know superconductors[1] for which superconductivity vanishes only in a field of around 600 kGauss (!). It is true that even for the best existing superconducting magnets the maximum field is still a few times smaller. However, this fact does not change at all the conclusion that the technical use of superconductors has become a reality and proceeds on an every increasing scale (see, for instance, Refs. 2 and 3). If it is necessary, as is now usually assumed, to use superconducting magnets for the creation of thermonuclear reactors with magnetic confinement, the importance of these magnets can in the future increase very strongly. However, an incommensurably larger effect could occur, if we obtained high-temperature superconductors.

The history of the study of superconductors is permeated by the aspiration to obtain materials with even higher values of the critical temperature T_c. The known situation about what has been achieved in this respect can be seen from Table 2.1. We show in Table 2.2, for reference, the boiling and melting temperatures (at atmospheric pressure) for a number of materials.

TABLE 2.1

Material	Critical Temperature T_c (°K)	Year of Discovery of its Superconductivity
Hg	4.1	1911
Pb	7.2	1913
Nb	9.2	1930
Nb_3Sn	18.1	1954
$Nb_3(Al_{0.75}Ge_{0.25})$	20 to 21	1966
Nb_3Ga	20.3	1971
Nb_3Ge	23.2	1973

TABLE 2.2

Substance	He	H_2	Ne	N_2	O_2	H_2O
Boiling temperature T_b (°K; under atmospheric pressure)	4.2	20.3	27.2	77.4	90.2	373.16
Melting temperature T_m (°K: under atmospheric pressure)	-	14.0	24.5	63.3	54.7	273.16

A comparison of Tables 2.1 and 2.2 indicates the fact that to cool known superconductors one can, apart from liquid helium, use only liquid hydrogen, but in theory rather than in actual practice. Firstly, it is most convenient to operate under atmospheric pressure, but in that case only Nb_3Ge remains superconducting in liquid hydrogen. (Moreover, as far as we know this material can not yet be used for the coils of electromagnets.) Secondly, both for obtaining high critical fields and currents, and from considerations about the reliability of the operation of the magnets, it is necessary to get rather far from the critical temperature. Therefore it is not realistic at the present time to use liquid hydrogen rather widely in superconducting devices. However, at the same time the era of 'hydrogen' or 'average-temperature' superconductivity seems to be close. To reach this it is sufficient to raise T_c to 27 to 30 K for a material which is suitable for applications to electromagnets. The efforts of the past, which are well reflected in Table 2.1, and also various estimates and considerations (see, in particular, what follows) give us grounds to hope that this problem can be solved in the traditional way of preparing and processing new alloys and compounds (in that connection we may mention, for instance, Nb_3Si).

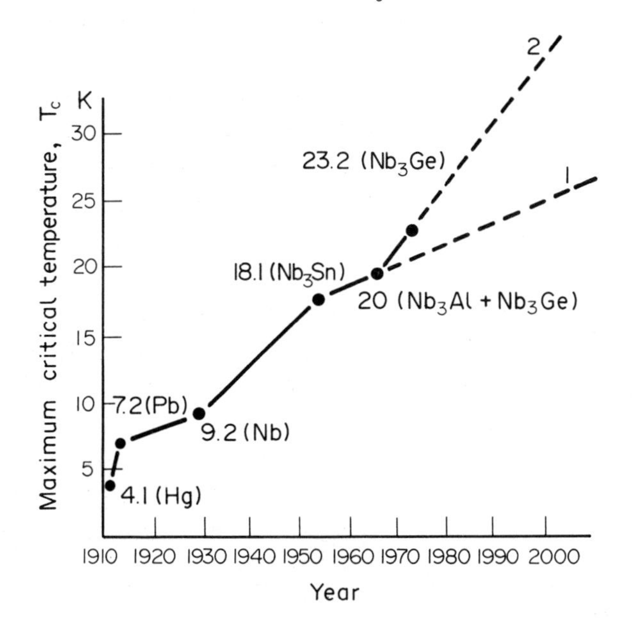

Fig. 2.2

When will this happen? It is impossible, I feel, to give here a well-founded answer. The only thing which I can conjecture, as a joke, is to use extrapolations in the graph of $T_c(t)$ which is given in Fig. 2.2. The first dashed line which one could have drawn already ten years ago 'predicts' that a temperature of ~30 K will only be reached in the next century. A more 'up-to-date' second straight line approaches that stage approximately in 1990. However, I repeat that all this talk about dates is not serious whereas the expectation of some growth itself is well founded. But up to what limits? And, in general, what determines the achievable values of T_c? and would it be impossible to produce 'high-temperature' superconductors for which the temperature T_c would, say, reach 90 K or even room temperature?

Superconductors with $T_c \geqslant 90$ K could be cooled by liquid nitrogen and, undoubtedly, this would lead (to be precise, when certain additional conditions are satisfied) to a genuine revolution in electrical technology and energetics. Of course, the production of materials which would remain superconducting at even higher temperatures would, in principle, lead to an even larger effect. However, would all this be possible? To answer this question means to solve the problem of high-temperature superconductivity (to be more precise, if the answer were negative one should talk about 'closing' this problem). We shall in what follows briefly consider the factors which determine the value of T_c and then we shall discuss ideas and means which enable us to hope for the creation of high-temperature superconductors (this problem is elucidated in more detail in a monograph edited by Ginzburg and Kirzhnits [4]; see also a review article by me[5] and a popular exposition[6]).

II. WHAT DETERMINES THE VALUE OF THE CRITICAL TEMPERATURE T_c?

More than 40 years after the discovery of superconductivity, practically up to 1957 the nature itself of this effect remained mysterious. It is true that after the discovery in 1938 of the superfluidity of liquid helium one could say that superconductivity is like the superfluidity (flow without friction) of the electrons in a metal. However, this was not yet a genuine explanation. Indeed, for temperatures $T > T_\lambda = 2.17$ K liquid helium, the ^{4}He atoms of which have spin zero, shows superfluidity. However, such particles (and in general particles with integer spin — bosons) undergo a Bose-Einstein condensation — when the temperature is lowered they 'accumulate' in the lowest energy level. Such a system (for instance, a real boson gas or, as is shown experimentally, liquid ^{4}He) exhibits superfluidity. However, in the case of liquid helium consisting of atoms of the lighter isotope ^{3}He, which have half-odd-integer spin, superfluidity does not set in until one reaches a temperature which is lower by a factor 1000 than for ^{4}He (this fact is immediately a clear demonstration of the rôle of the Bose-Einstein statistics and condensation from the point of view of the appearance of superfluidity). On the other hand, electrons, like ^{3}He atoms, have spin $\frac{1}{2}$ and therefore, it would appear, should not form a superfluid system, which the presence of charges turns into a superconducting gas or liquid. Nonetheless, superconductivity turned out to be, indeed, related to superfluidity not only in its manifestation, but also in the nature of the phenomenon. In fact, superconductivity begins in a metal when the conduction electrons with opposite spins and momenta attract one another. This attraction leads to the formation of pairs — quasi-atoms consisting of two electrons, evidently, with total spin zero.* The pairs are thus bosons and can undergo a Bose-Einstein condensation. The situation here is, in principle, analogous to the one occurring in the case of ^{4}He atoms — they consist, in fact of a nucleus (two protons and two neutrons) and two orbital electrons. However, protons, neutrons, and electrons have spin $\frac{1}{2}$ and only their combination — the ^{4}He atom has spin zero. On the other hand there is also a large difference between superconductors and liquid ^{4}He. The point is that the dimensions of the electron pairs (a typical radius is of the order of 10^{-4} cm) are considerably larger than the average distance between the electrons. The electron pairs therefore overlap appreciably and lose their individuality, and there arises some unique 'coherent' state of the electron system. What we have said is, nonetheless, the basis of the physical understanding of superconductivity; we have, of course, no possibility to discuss this here in detail.

In order to break up (split) an electron pair at the absolute zero (at $T \equiv 0$) we need to spend some energy† $2\Delta(0)$. When the temperature increases this break-up energy $2\Delta(T)$ decreases and at $T \equiv T_c$ it vanishes, that is $\Delta(T_c) = 0$. It is rather natural therefore that the critical temperature, at least in the simplest cases is such that $T_c \sim \Delta(0)/k$, where $k = 1.38 \times 10^{-16}$ erg/degree is the Boltzmann constant. The gap $\Delta(0)$ and thus also T_c is the larger the stronger the conduction electrons stuck together in pairs attract one another. However, it is impossible to characterize this 'attractive force' by a single parameter, one must use at least two, The fact is that in a normal (not superconducting) metal the conduction electrons are distributed in energy such that they occupy levels with energies from zero up to the

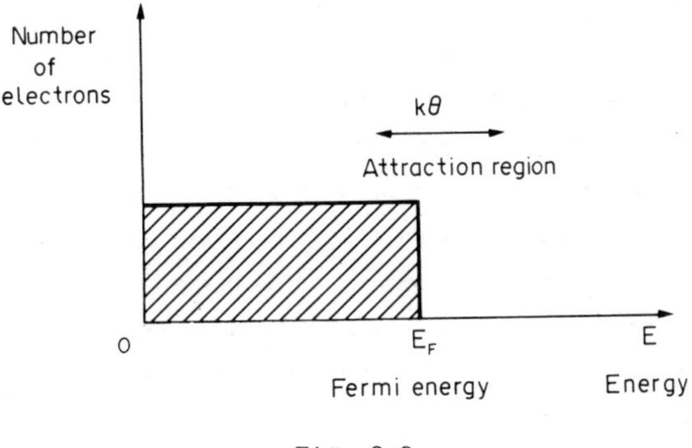

Fig. 2.3

* In principle. it is also possible that pairs with spin unity are formed which also are bosons. However, for the known superconductors this possibility is not realized (in the case of neutrons and protons in neutron stars, and also for ^{3}He the situation is different).

† In this connection one speaks about a gap in the energy spectrum, and to release a single electron we need spend an energy $\Delta(0)$.

Fermi energy E_F (this distribution for a free electron gas at $T=0$ is shown in Fig. 2.3). Such kind of distribution is stable and is retained if the electrons repel one another – and this occurs in normal metals. As we have already stated, superconductivity arises when there is attraction between the electrons, but now we must be more precise and state that one is dealing only with attraction in some (energy) neighbourhood near the Fermi energy E_F. In other words, it is necessary that the electrons attract one another in some energy layer with width $k\theta$ where θ is an appropriate temperature (see Fig. 2.3). This temperature θ is one of the parameters which determines the value of T_c in the simplest model of a superconductor – the Bardeen-Cooper-Schrieffer (BCS) model. The second parameter is a dimensionless constant λ_{eff} which characterizes the magnitude of the attraction.

The critical temperature is then given by the BCS formula (1957)

$$T_c \sim \theta e^{-1/\lambda_{eff}} . \qquad (i)$$

By virtue of what we said earlier the superconducting gap in this model is

$$\Delta(0) \sim kT_c \sim k\theta e^{-1/\lambda_{eff}} .$$

Before we go any further we must note that we have postulated the presence of an attraction between some of the electrons. The possibility of an attraction between electrons itself may cause surprise as it is well known to everybody that similar charges repel one another. This statement, indeed, is unshakably true, if we are dealing with charges in vacuo. However, the conduction electrons are in a metal where there are ions apart from the electrons; these form a crystalline lattice. As a result the interaction energy V between any two conduction electrons which we consider is radically changed and, roughly speaking, consists of two parts: $V = V_C + V_a$. Here V_C is the energy of the Coulomb interaction between the given electrons (it is positive, which corresponds to repulsion; in vacuo $V_C = e^2/r$, where r is the distance between the charges e, but in a metal the energy V_C decreases strongly with increasing r as the result of the screening of the electron considered by all the other conduction electrons). The part V_a of the interaction energy takes into account the contribution from the lattice, and also from all 'bound' electrons which are not part of the system of conduction electrons. The energy V_a can be negative (attraction). Moreover, the contribution from the lattice is in the range of electron energies which is important for superconductivity always negative, and the rôle of the 'bound' electrons in normal metals is in the majority of cases small.

That there is an attraction between the conduction electrons caused through their interaction with the lattice can be shown and explained both in classical and in quantum mechanical language. In the latter case one talks about the fact that two interacting electrons exchange phonons – the energy quanta of the lattice oscillations. The shortest wavelength in the lattice $\Lambda_{ph,min} \sim 3 \times 10^{-8}$ cm, that is, of the order of the lattice constant. The maximum frequency of the lattice vibrations $\omega_{ph,max} \sim 2\pi u/\Lambda_{ph,min} \sim 10^{13}\ s^{-1}$, as the sound speed in a metal $u \sim 10^5$ cm/s. Hence, the maximum phonon energy is $\hbar\omega_{ph,max} \sim 10^{-14}$ erg ~ 0.01 eV.

In a very rough approximation the dimensionless interaction constant $\lambda_{eff} = \lambda - \mu$, where the constants λ and μ are proportional to $|V_a|$ and V_C, respectively. In this approximation superconductivity occurs, if the attraction (λ) caused by the rôle of the lattice overcomes the Coulomb repulsion (μ). However, in actual fact, because of a different frequency (energy) dependence of these two interactions, the Coulomb repulsion turns out to be 'suppressed' as compared to the attraction λ so that

$$\lambda_{eff} = \lambda - \mu^* , \quad \mu^* = \frac{\mu}{1 + \mu \ln(\theta_F/\theta)} = \frac{\mu}{1 + \mu \ln(\omega_F/\omega_c)} , \tag{2}$$

where ω_F and ω_c are frequencies which correspond to the Fermi energy ($E_F = k\omega_F = \hbar\omega_F$) and to the region near the Fermi surface in which the attraction acts ($k\theta = \hbar\omega_c$).

When the deciding quantity for the attraction between the electrons is their interaction with the phonons, the frequency $\omega_c \sim \omega_{ph,max}$ for the phonon mechanism of superconductivity (clearly, phonons cannot transfer energies larger than $\hbar\omega_{ph,max}$). Therefore, the rôle of the temperature θ in the BCS formula (1) is in this case played by the Debye temperature θ_D which is just of the order $\theta_D \sim \hbar\omega_{ph,max}/k$. Moreover, in that case $\mu^* \ll \mu$, as the Fermi frequency in metals

$$\omega_F = E_F/\hbar \sim 10^{15} \text{ to } 10^{16}\ s^{-1} \quad (E_f \sim 1 \text{ to } 10 \text{ eV}),$$

so that

$$\omega_F/\omega_{ph,max} \sim 10^2 \text{ to } 10^3 \text{ and } \ln(\omega_F/\omega_{ph,max}) \sim 5 \text{ to } 10 .$$

The inequality $\lambda_{eff} > 0$ (see (2)) which is necessary for superconductivity is thus satisfied rather often and many metals (including alloys and compounds) turn out, indeed, to be superconductors.

In order to be clear I repeat the important statement: when we have a phonon mechanism for superconductivity so that this phenomenon is caused by the interaction of the electrons with the lattice, the role of θ in the BCS formula (1) is played by the Debye temperature θ_D. Usually, $\theta_D \sim 100$ to 500 K, as is known from specific heat measurements and other data. As far as the interaction constant λ_{eff} is concerned, which must be positive for a superconductor, in the case of a phonon mechanism the 'suppression' of the Coulomb repulsion (replacement of μ by μ^*; see (2)) is very favourable. Usually, $\lambda_{\mathbf{eff}} \leqslant 1/3$, and this means that even when $\theta_D \sim 500\,\mathrm{K}$, the critical temperature $T_c \leqslant 500 \times e^{-3} \sim 25\,\mathrm{K}$. From this it is already qualitatively clear why the phonon mechanism cannot lead to high-temperature superconductivity.

I have used here the BCS formula (1) and also attempted to elucidate a few points dealing with the superconductivity mechanism in general and with the phonon mechanism in particular. However, I should like to be more definite and note that such a level of exposition does not correspond to the state of the theory of superconductivity at the present time. The appearance of the BCS theory of 1957 was for the micro-theory of superconductivity a watershed which opened the way to a colossal number of papers which developed, refined, and generalized the initial variant of the theory. I did not take part in this process, as I do not possess the necessary complicated technique needed for solid state theory, and also somehow this kind of activity does not suit me. Therefore, when with the birth of the BCS theory the mantle of mystery was lifted from the phenomenon of superconductivity, I thought that, in general, I would leave the superconductivity business (I started working in this field about forty years ago — my first paper was published in 1944, but as far as results were concerned I occupied myself basically with a phenomenological or, to be more precise, a semi-phenomenological theory). I made this auto-biographical kind of remark (and I hope that it turns out to be justified) for several reasons. In actual fact, in the field of superconductivity one could find also after 1957 a large number of interesting problems for the solution and discussion of which at a certain level one does not need to use on the whole complicated mathematical methods. I can mention here, bearing in mind my own research, superconductivity and superfluidity in the cosmos (first of all I have neutron stars in mind), thermo-electric effects in superconductors, and the problem of high-temperature superconductivity. The study of the superfluidity of helium II near the λ-point and some other problems about superfluidity are close to these problems. I shall report on some of my own work in the field of thermo-effects and superfluidity in a more specialized seminar.

Returning to the problem of high-temperature superconductivity, in which I became interested in 1964, I wish to discuss how the corresponding research developed. At first optimism reigned, and it seemed that qualitative or semi-quantitative considerations indicated promising ways for our search. At this level I wrote, in particular, some review papers [5,6] and at just this level I shall give the exposition in this lecture. However, already ten years ago, yes, strictly speaking, even somewhat earlier, it became sufficiently clear that for a study of the problem of high-temperature superconductivity it was necessary to pull in all possibilities which there are in the theory and to analyze the problem from different sides, including also the 'highly theoretical' level. To do this we created a special group in the I.E. Tamm Theoretical Physics Department of the P.N. Lebedev Physical Institute of the USSR Academy of Sciences (I became head of that division in 1971 when its founder I.E. Tamm died) and we organized a special weekly seminar. The authors of the monograph[4] which we wrote with the aim to elucidate the state of the problem (of course, in a book published in 1977 we could only report on the state of the art of around 1976) form the core of this group. Since then research has, of course, been continued; recently the most important new results have been published in two review papers[7], but also some other papers have appeared and there have been some new directions of research (the problem of superdiamagnetics which I shall mention in what follows).

The present state and, so to speak, evolution of ideas in the field of study of high-temperature superconductivity in some respects reminds one of the situation in the field of producing devices for controlled thermonuclear fusion. As I mentioned in one of the earlier lectures, thirty years ago when the problem of controlled fusion had just been recognized (one might say, had been born) it seemed that it would not be all that difficult to solve it. To be more precise, one assumed that the difficulties to be encountered would rather be technological, engineering ones. The idea itself of toroidal magnetice traps appeared sufficiently reliable to guarantee success. However, we now know that soon afterwards enormous and not at all 'technological' complications were met with which were connected first of all with instabilities characteristic for a collisionless plasma, and also with a whole range of secondary effects. It became clear that the problem of controlled fusion would not be solved without a profound and all-sided development of plasma physics. This process is continuing up to now although, apparently, at the present time many problems have already been left behind, and, at least for tokamaks, the stage is now occupied, mainly, by engineering problems (however, I am not completely confident that these estimates are correct). As far as the problem of high-temperature superconductivity is concerned, recently it also became very clear both on the basis of accumulated experience of rather unsuccessful attempts to raise T_c and as the result of developing the theory itself that it is necessary to have a much deeper understanding of the factors which determine the critical temperature T_c in real superconducting systems (both already existing ones and also ones one could think of).

However, this new stage in the field of the study of high-temperature superconductivity has only started, its outlines have not yet been sharply traced. Therefore, I can in the present lecture only give a few appropriate indications in section 4. In the next section we proceed along the same path as in the present one — we remain on the level of qualitative discussions and simple estimates, and we also give some results obtained using approximations which are more or less normal for contemporary superconductivity theory — that is, without taking into account the effects of the local field and a few other effects; see Ref. 7 and section 4.

III. WAYS OF PRODUCING HIGH-TEMPERATURE SUPERCONDUCTORS

In the previous section we gave the simplest considerations which enabled us to understand, so to say, the limitations of the possibilities of the phonon mechanism of superconductivity on the road to enhancing the critical temperature T_c. In actual fact, one can do appreciably better (see Ref.4 and the literature cited there). It is impossible to use the BCS formula (1) in the case of most interest, as it is itself true only in the so-called weak coupling case, when $\lambda_{eff} \ll 1$. In the general case, however, we get a formula of the type

$$T_c \sim \theta\, e^{-(1+\lambda)/(\lambda-\mu^*)} \ . \tag{3}$$

Of course, if $\lambda \ll 1$, (3) becomes (1).

In the case of the phonon interaction it turns out that the interaction constant λ decreases with increasing θ_D, and, roughly speaking $\lambda \propto \theta_D^{-2}$. It is therefore altogether impossible to increase T_c strongly, even if we take an increase in θ_D into account. Further, when we evaluate T_c for the case of a phonon mechanism we must know the whole spectrum of the lattice vibrations and also a whole range of other data about the metal. The corresponding calculations substantiate the conclusion that for known metals and alloys $T_{c,max} \leqslant 25$ to 40 K, when we use the phonon mechanism. This was just the estimate I had in mind earlier when I said that there are still some reserves on the road to producing new superconductors of the 'normal' kind.

And what is the situation regarding hypothetical 'unusual', high-temperature superconductors? On the one hand, one has here some hope to use the same phonon attraction mechanism, but under special conditions. On the other hand, one can hope to apply non-phonon mechanisms which can, in principle, lead to superconductivity.

We start with the first kind of possibilities and, to be concrete, with metallic hydrogen or deuterium. The Debye temperature for metallic hydrogen is $\theta_D \sim 3000\,K$, which is connected with the small mass of the nuclei. Moreover, in this case, which is exceptional (there are, clearly, no bound electrons in metallic hydrogen) the constant λ remains, apparently, not too small. As a result it is quite well possible that for metallic hydrogen $T_c \sim 100$ to $200\,K$. However, metallic hydrogen has not only not yet been produced, but it is unknown whether it remains for any length of time in the metallic, even if metastable, state when the pressure is lifted. It is thus hardly judicious to connect the problem of high-temperature superconductivity with the other interesting problem or, if you please, dream – the production and study of metallic hydrogen.

The known expectations for the enhancement of T_c can be connected also with obtaining superconducting hydrogen-containing substances (candidates are, for example, LiH_2F and alloys based upon PdH under sufficiently high pressures; in principle, in the case when the appropriate metastable phase exists superconductivity could be retained also after the pressure is lifted). Another, more attractive possibility is the production of metals containing light atoms. For instance, the compound $(SN)_x$ which contains not a single metallic atom (!) is a metal at low temperatures. Moreover, this compound is superconducting – although the critical temperature is small, it lies around 0.3 K. Relatively recently one began to hope also that it might be possible to produce 'organic' metals, consisting of organic molecules. The presence of the light carbon atom (C) and the probable presence of hydrogen guarantees in such compounds a large effective value for the Debye temperature θ_D or its analogues (in fact, we are talking about the high frequencies of the oscillations in the lattice). On the other hand, in 'organic' metals one can expect the Coulomb interaction to be relatively small due to the large size of the molecules. As a result one may hope to obtain substances with increased values of T_c. Unfortunately, the problem of the properties of 'organic' metals and of other metals containing light atoms and possessing appropriate properties (in particular, large values of T_c) still remains insufficiently clear both in experimental and in theoretical respects. It is true that in 1980 there was the first success. Organic and at the same time superconducting metals have already been synthesized.[8] The corresponding values of T_c are small and it is unclear whether there are any possibilities to raise T_c. However, there are doubtlessly perspectives in that direction.

The second and, probably, a wider field for the search for high-temperature superconductors is connected with the use of new, non-phonon superconductivity mechanisms. The point is that the attraction between the conduction electrons which is necessary for the occurrence of superconductivity can arise, apart from lattice (phonon) effects, also due to effects due to the other ('bound') electrons or to some substances (molecules, dielectric) being in contact with them. Essentially we are thinking here always of a contribution from various constituents parts of the substance – the lattice of the nuclei (ions) or of the 'bound' electrons – to its dielectric permittivity (it is just the frequency- and wavelength-dependence of this permittivity which determines the interaction between the electrons). Superconductivity 'thanks to' the electron part of the metal could be said to be caused by an electron mechanism. However, there would arise here some confusion as in a metal in the case of any kind of superconductivity — which afterall is an electronic phenomenon – the superconductivity is due to the conduction electrons. We therefore call this mechanism an exciton mechanism although this terminology is not always exact. However, in a number of cases it is not only valid, but even descriptive.

Indeed, generally speaking, in a solid there may propagate not only sound waves (phonons), but also other kinds of excitations, the existence itself and the characteristics of which (such as the frequency and speed) are determined by the electrons and not by the ions (lattice). Such excitations are often called electron excitons or simply excitons although often one also applies other names. For instance, in solids one can have the propagation of electron excitons called plasmons and which are the complete analogues of the longitudinal waves in a plasma(the characteristic frequency of these waves is

$$\omega_e \sim \omega_p \sim \left(\frac{4\pi e^2 n}{m}\right)^{\frac{1}{2}} = 5.64 \times 10^4 \, n^{\frac{1}{2}} \, s^{-1}$$

where n is the electron density and e and m the charge and mass of an electron). In general, longitudinal excitons, in which the electrons oscillate in the direction of the wave propagation, appear always when the dielectric permittivity of the medium ε is equal to zero. For a plasma under well defined conditions (in particularly, for long wavelengths) $\varepsilon = 1 - (\omega_p/\omega)^2$ and the equation $\varepsilon(\omega) = 0$ just gives the plasmon frequency ω_p.

The exchange of excitons, similar to the exchange of phonons can lead to attraction between the conduction electrons. In that case the rôle of the Debye temperature $\theta_D \sim \hbar\omega_{ph,max}/k$ is now played by the temperature $\theta_e \sim \hbar\omega_e/k$, where ω_e is the exciton frequency which can easily reach very high values of the order of $\omega_F \sim 10^{15}$ to $10^{16}\,s^{-1}$ (for instance, the plasma frequency in metals $\omega_p \sim 10^{15}$ to $10^{16}\,s^{-1}$, as $n \sim (1$ to $3) \times 10^{22}\,cm^{-3}$). Such 'energetic' excitons are not useful from the point of view of increasing T_c, as $\ln(\omega_F/\omega_e)$

is in this case small and thus $\mu^* = \mu$ (see (2)). However when

$$\omega_e \sim (1 \text{ to } 3) \times 10^{14} \text{ s}^{-1},$$

that is,

$$\hbar\omega_e \sim 0.1 \text{ to } 0.3 \text{ eV},$$

the temperature $\theta \sim \theta_e \sim 1000$ to 3000 K and there is thus for the exciton mechanism at any rate no limiting effect from the smallness of the temperature θ in equations (1) and (3) on obtaining high values of T_c.* However, even when θ_e is large one can obtain a high value of T_c only for a sufficiently strong interaction between the electrons and the excitons, that is, when $\lambda_{eff} \geqslant \frac{1}{3}$ to $\frac{1}{4}$.

The main problem is whether such values are attainable. Besides, there is the not less important problem: can excitons of the required type propagate in a metal? Indeed, sound (phonons) can propagate in any body. It is true that the value of T_c for the phonon mechanism of superconductivity is determined by sound waves with the shortest wavelength — usually with wavelengths of the order of the lattice constant (this means, roughly speaking, that the momentum of those phonons is of the order of the momentum p_F of the conduction electrons at the Fermi surface), but just those phonons are very important in a solid and we know that they play their part in normal superconductors. As far as excitons are concerned, however, the situation is not so clear and the problem consists just in indicating favourable conditions for the action of the exciton mechanism.

It is clear from what I have said that for this we need, generally speaking, to have excitons (or, as it is usually stated, an exciton band) in the material in the energy range $E_e = \hbar\omega_e \sim 0.1$ to 0.3 eV. However, as far as I know, there are no such excitons in a good metal. This is in general understandable: we have already mentioned that excitons with a

---

* The hope of raising T_c as the result of the action of an 'electron' or exciton mechanism can be explained also in another way. The parameter θ in formula (1) and (3) is in the simplest models connected through the relation $\theta \propto M^{-\frac{1}{2}}$ with the mass M of those particles who are the 'intermediaries' for producing a dominating attraction between the conduction electrons. For the phonon mechanism the ion mass plays the role of M and, indeed, the Debye temperature $\theta_D \propto M^{-\frac{1}{2}}$. Hence it is clear (see (1) and (3) with $\theta \sim \theta_D$), in particular, that the isotope effect in T_c the discovery of which in 1950 played a clear rôle in understanding the nature of 'normal' superconductivity, is caused by the action of the phonon mechanism. However, in the case of the exciton mechanism the electrons (which are different from the conduction electrons) are the 'intermediaries' and the electron mass m plays the rôle of the mass M, so that

$$\theta \sim \theta_e \sim \theta_D (M/m)^{\frac{1}{2}} \sim 10^2\, \theta_D \leqslant 10^4 \text{ to } 10^5,$$

as we mentioned earlier.

frequency ω_e appear then when in that frequency range the dielectric permittivity of the material $\varepsilon(\omega) = 0$ while in vacuo $\varepsilon = 1$. Thus, the medium must strongly alter ε and under the conditions which we are interested in this is possible only when there are 'bound' electrons present with a binding energy of the same order of magnitude $E_e \sim 0.1$ to 0.3 eV. However, in a good metal where there are conduction electrons with energies up to the Fermi energy $E_F \sim 1$ to 10 eV it is very difficult to keep weakly bound electrons. In general, the occurrence of the exciton mechanism in a good (normal) metal is very difficult and very unlikely.

However, this conclusion has not yet been sufficiently substantiated. Some authors assume that there is a possibility to realize the exciton mechanism already in three-dimensional systems (metals) of a more or less ordinary type. Moreover, in some metals (for a metal with almost coinciding electron and hole Fermi surfaces or for a metal with narrow allowed bands) structural and superconductive transitions may coexist and 'interfere' and this must under certain conditions lead to an appreciable increase in T_c. In this case the attraction between the conduction electrons may be caused by the phonon mechanism, but the structural transition is of an electron nature and its temperature being close to T_c leads to an increase in the density of states near the Fermi surface and thereby to an increase in T_c. Under these conditions the term 'exciton mechanism of superconductivity' is, of course, arbitrary, but it is also difficult to speak of a phonon mechanism (see Ref. 4, Chapter 5).

We now turn to systems which attracted attention in 1964 and thereby opened up the present stage of the discussion of the problem of high-temperature superconductivity (for references see Refs. 4 to 6). We are referring here to metallic 'chains' or filaments with polarizers distributed on the side, to filamentary compounds, dielectric-metal-dielectric 'sandwiches', layer compounds, and so on. In all these cases the basic idea is to combine a good conductor (metal) in which there are no suitable excitons with a dielectric part (molecules, dielectric coatings or layers) which have the necessary exciton

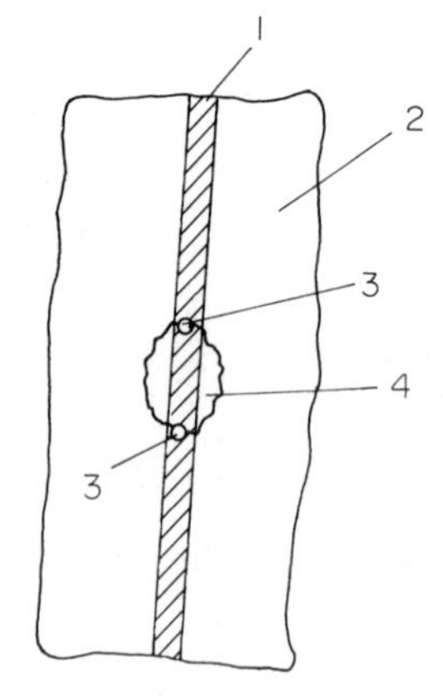

1 - Metal; 2 - Non-metal (dielectric)
3 - Electron; 4 - Exciton

Fig. 2.4

spectrum (see Fig. 2.4 which refers directly to a dielectric-metal-dielectric 'sandwich'). It is true that such a division into a metallic and a dielectric system is possible only, really, when we talk about it, rather than when we are making it. The reason is that the excitons are quickly damped in the metal and altogether operate only in a thin layer of atomic dimensions at the boundary of the metal with the dielectric. This means that in the case of a 'sandwich' the metallic film must have a thickness of the order of or less than 10 to 20 Ångström. To produce such a 'sandwich', especially with suitable dielectric coating, is difficult and on the whole is as yet an unsolved problem. The situation is better with respect to layer compounds which are, so to say, natural stacks of 'sandwiches'. Metallic conductivity of layers of a 'metal' is secured in a number of such compounds. Moreover, the dielectric 'layers' can be changed. This is, for instance, achieved by 'intercalation' — incorporating different molecules between layers of metal. In this way a whole new class of superconductors was discovered. It was proved in this way that there was the possibility that practically two-dimensional superconductors existed. At the same time, however, no dramatic increase in T_c was obtained, but this could be explained completely by the nature of the intercalated organic molecules which, in particular, did not have sufficiently low excited electron levels. In order that the 'dielectric' layers in layer compounds will have the properties which are necessary for obtaining high values of T_c it is desirable to make them not from large molecules but from some kind of collectivized semiconductor.

The study of various layer superconductor materials seems to me to be one of the directions of further investigations which has perspectives. Incidentally, this is so even independent of the problem of raising the critical temperature (it is sufficient to indicate that some layer compounds have exceptionally high critical magnetic fields parallel to the layer; superconducting layer compounds are also of interest in connection with some other peculiarities they have). In the case of synthetic dielectric-metal-dielectric 'sandwiches' some perspectives also may be opened up for obtaining rather high values of T_c. We must here emphasize that the study of 'sandwiches' is closely connected with the study of the dividing boundaries and of surface layers, the properties of which are not well known in a number of respects (moreover, several possibilities are opened up here). In general, in the light of the vigorous development and the successes of surface physics, the study of superconductors which are connected with surfaces acquires a special value (this includes 'true' surface superconductivity in surface levels[9,10] and the study of the effect on T_c of chemical compounds and chemisorption on surfaces).

Unfortunately, it is as yet impossible to give a reliable calculation of T_c in the case of the exciton mechanism even for quasi-homogeneous materials (without speaking about strongly non-uniform structures like 'sandwiches'). The absence of the necessary data about the permittivity ε of the corresponding substances in a wide frequency and wavelength region is particularly troublesome in this case. As far as general considerations and estimates, based upon an application of formulae of the kind (3) with θ replaced by θ_e, are concerned, they are in no way inconsistent with the possibility of reaching values of $T_c \leqslant 100$ to 300 K. It is a different situation with respect of the question whether we may discover in the future some circumstances by virtue of which the constant for the coupling between the excitons and the electrons, λ, cannot be sufficiently large (say, reach values of the order of 1/3 or 1). However, at this moment there are as yet no indications that it is impossible to guarantee sufficiently strong electron-exciton interactions provided we satisfy possibly very stringent and special, but altogether in principle, achievable, conditions (see also section 4 below).

Apart from two-dimensional and quasi-two-dimensional systems, one-dimensional and quasi-one-dimensional systems have also attracted attention, as we have already mentioned. It is true that along this road for a long time it was impossible to make progress experimentally, simply because there did not exist sufficiently long molecular 'chains' with metallic conductivity. However, rather recently, a vigorous study has started of quasi-one-dimensional conductors, firstly, of the type $K_2Pt(CN)_4Br_{0.3}3H_2O$ (abbreviated as KCP) and tetrathiafulvalene-tetracyanoquinodimethane (TTF-TCNQ). It was ascertained, however, that such materials change, to a non-conducting state (more precisely, become semiconductors) when the temperature is lowered, if they have a rather well pronounced quasi-one-dimensional nature (that is, if the coupling between neighbouring chains is weak). However, $(SN)_x$, several substances with a TCNQ base, and other substances remain metals at arbitrarily low temperatures, but in those cases we are dealing with materials which are in some respects very far from being one-dimensional (rather, they can be ranked among the strongly anisotropic three-dimensional metallic structures). Obtaining high-temperature superconductors on the basis of a clearly pronounced quasi-one-dimensional structure (even forgetting strictly one-dimensional chains) is therefore at this moment unlikely. At the same time, the wide range which the study of quasi-one-dimensional structures has occupied recently is very symptomatic and indicative. It is now probable that everybody understands how interesting and full of perspectives the study of new kinds of conductors, such as quasi-one-dimensional or layer conductors, is. Meanwhile it is not so very long ago that it was thought to be hardly a sign of good taste to make ironical remarks about the attempts to synthetize 'organic' and, in general, high-temperature superconductors.

When discussing the problem of high-temperature superconductivity it is nowadays impossible not to mention also the diamagnetic anomalies observed in 1978 in CuCl and in 1980 in CdS (see Refs. 10, 11, and the literature given there). In those substances with impurities, and only under certain conditions and for certain methods of preparation even at liquid nitrogen temperatures super-diamagnetism has been observed, that is, the magnetic susceptibility χ was comparable with the susceptibility of an ideal diamagnetic $\chi_{id} = -1/4\pi$ (in all diamagnetic which were known earlier $|\chi| \ll |\chi_{id}|$ and usually $|\chi| < 10^{-4}$). This is just how superconductors behave in a magnetic respect (to be more precise, the complete Meissner effect which can be observed in superconductors corresponds just to the value $\chi = \chi_{id}$). However, there is as yet no guarantee that the observed phenomena are not due in some way to subsidiary effects. Moreover, it is possible, in principle, that super-diamagnetism is not connected with superconductivity.* On the other hand, it seems to me to be totally permissible to guess that one has already observed high-temperature superconductivity in CuCl and CdS. This can only be ascertained as the result of further experiments.

We still must mention strongly non-equilibrium superconducting systems which can be produced, for instance, be laser 'pumping' of non-equilibrium

---

* We are talking here about semiconductors (that is, solids which do not have metallic conductivity, let alone superconductivity) which have anomalously large diamagnetic susceptibility $|\chi| \leq |\chi_{id}|$. Such a susceptibility may occur (see Refs. 12 and 10, and the literature quoted there) by virtue of the existence in the sample of closed spontaneous currents with large characteristic dimensions.

† We emphasize, in general, that up to recently superconductivity was studied in equilibrium conditions (in the state of thermodynamic equilibrium or for quasi-equilibrium metastable phases). At the same time superconducting properties can, of course, be maintained also in non-equilibrium conditions These occur in very many greatly varying forms, as for a given metal there may be differences not only in such parameters as the temperature, but also in the form of the electron and phonon distribution functions. It is therefore clear that when we go over to a study of non-equilibrium superconducting states a very broad field of activities is opened up. One can think that in the near future the main direction in the field of superconductivity physics will not only be concerned with the study of new substances (in particular, with the aim of producing high-temperature superconductors), but also just with the investigation of superconductivity under non-equilibrium conditions.

electrons in metals or semiconductors. Research in this direction has started not very long ago and it has, at the same time, very broad possibilities for reaching high values of T_c (see Ref. 4).[†]

IV. IS THERE A REAL POSSIBILITY TO INCREASE T_c USING THE EXCITON MECHANISM?

In the foregoing we connected the possibility of producing high-temperature superconductors ($T_c \gtrsim 100$ K) in first instance with the exciton mechanism of superconductivity (we mentioned also some other possibilities but we leave those now to one side). Using the BCS formula (1),(2) (or, more exactly, in terms of the BCS theory) one sees easily that the factor in front of the exponent $\theta = \theta_e$ can for an exciton mechanism be appreciably larger than the Debye temperature θ_D which occurs in (1),(3) in the place of θ for the case of the phonon mechanism. However, in fact there is in formulae (1), (3) also the exponential factor (the factor $\exp(-1/\lambda_{eff})$ in (1)). In the case of the exciton mechanism it is sufficient for the inequality $\lambda_{eff} < 1/5$ to be satisfied for $T_c \sim \theta_e \exp(-1/\lambda_{eff}) < 20$ K, when $\theta_e \sim 3000$ K. From this it is clear — as we emphasized also earlier — that it is impossible to give any guarantee for the effectiveness of the exciton mechanism from the point of view of reaching a large T_c. We need here a deeper analysis. Such an analysis has not at all been performed properly and in the present lecture we cannot in detail present what has already been done[4,7]. Nonetheless we make a few remarks in the present section at a somewhat different level than we did earlier (if they wish to, the readers can omit the present section; its results are taken into account in the next, concluding section).

Turning to formula (2) we see that when θ increases (as we said, for the exciton mechanism $\theta = \theta_e$), the quantity μ^* increases and, hence, λ_{eff} decreases. If we restrict ourselves for the sake of simplicity to the weak coupling case ($\lambda_{eff} \ll 1$) we return to the formula

$$T_c = \theta_e \exp\left[1/\left\{(\mu/[1+\mu \ln(\theta_F/\theta_e)]) - \lambda\right\}\right]$$

which can be obtained from (1) and (2). Equating the derivative $dT_c/d\theta_e$ to zero we see that the value of T_c as a function of θ_e has a maximum (we assume that $\theta_e < \theta_F$). The optimum value (corresponding to the maximum) is

$$\left.\begin{aligned} \theta_{e,opt} &= \theta_F \exp[(1/\mu) - (2/\lambda)], \\ T_{c,max} &= \theta_F \exp[(1/\mu) - (4/\lambda)] = \theta_{e,opt}\, e^{-2/\lambda} \end{aligned}\right\} \quad (4)$$

---

† **See footnote on previous page.**

(analogous formulae for the strong-coupling case of the kind (3) can also easily be obtained and are given in Ref. 4).

In the approximation of the superconductivity theory which is often used[4]

$$\mu - \lambda = N \left\langle \frac{4\pi e^2}{q^2 \varepsilon(0,q)} \right\rangle , \tag{5}$$

where N is the density of states at the Fermi boundary in the corresponding normal metal (state), $V = 4\pi e^2/q^2\varepsilon(\omega,q)$ is the energy of the interaction between two electrons in the Fourier representation (ω is the frequency, $\hbar\vec{q}$ the momentum, and by virtue of the assumed isotropy of the problem only the length of the wavevector $\vec{q}$, that is, the quantity q enters); the brackets $\langle\ \rangle$ in (5) correspond to an average over the phonon and the electron spectrum and also over the Fermi surface of the conduction electrons. Here $\varepsilon(\omega,q)$ is the complex dielectric permittivity of the medium for longitudinal waves (it is well known that when $q \neq 0$ the permittivity for longitudinal and transverse fields are different). On the whole, if we are not interested in local field corrections and the nature of the averaging, the appearance of the permittivity ε is total understandable (if ε does not depend on ω and q the Fourier transform $V(q) = 4\pi e^2/q^2\varepsilon$ corresponds to the interaction energy $V(r) = e^2/\varepsilon r$).

Finally, up to relatively recently the opinion was widespread that the stability condition of the system or the dispersion relations (the Kronig-Kramers relations) led to the requirement that

$$\varepsilon(0,q) > 0 . \tag{6}$$

If we accept the inequality (6), it follows from (5) that $\lambda < \mu$. However, even if $\lambda = \mu$, we have according to (4) $T_{c,max} = \theta_F e^{-3/\mu}$. It follows from various estimates and calculations that $\mu \leqslant 0.5$, and hence

$$T_{c,max} \leqslant \theta_F e^{-6} \sim 25\ K$$

for $\theta_F \sim 10^4$ K. When $\lambda = \mu = 0.5$ the strong-coupling approximation is more exact and we get $T_{c,max} = \theta_F e^{-10}$ and the value of T_c is yet lower, even when $\theta_F \sim 10^5\ K \sim 10$ eV.

One can thus think that the value of T_c is, in principle, limited and such a conclusion has been reached in the literature. However, in actual fact such a result is incorrect. Already from an interpretation of the experimental data for superconductors cases are known when $\lambda > 1$ and $\mu < 0.5$. Using the values $\lambda \sim 1$ and $\mu \sim 0.5$ leads on the basis of the strong-coupling formulae[4] to the estimate

$$T_{c,max} \leqslant 300\ K \quad , \quad \theta_{e,opt} \sim 10^3\ K . \tag{7}$$

To confirm this result one must, on the one hand, take into account the limited applicability of expression (5). However, most importantly, the main inequality (6) does not need to be satisfied at all.[4,7] The correct requirement of stability of the system and the satisfying of the dispersion relations has the form

$$\frac{1}{\varepsilon(0,q)} \leqslant 1 \quad , \quad q \neq 0 , \tag{8}$$

from which it follows that one of the following inequalities must be satisfied:

$$\varepsilon(0,q) \geqslant 1 \quad , \quad \varepsilon(0,q) < 0 \quad , \quad q \neq 0 . \tag{9}$$

Referring for details to Refs. 4 and 7 we restrict ourselves here to the remark that it is just the quantity $1/\varepsilon(0,q)$, and not $\varepsilon(0,q)$ which obeys the dispersion relations which are obtained from the requirement that causality is satisfied. The fact is that the electrical field $\vec{E} = \vec{D}/\varepsilon$, while the induction $\vec{D}$ can be 'controlled', as it is determined by the external charges (indeed, $\mathrm{div}\,\vec{D} = 4\pi\rho_{\mathbf{ext}}$). Therefore, $\vec{D}$ can be considered to be the 'cause', and the field $\vec{E}$ the 'effect'; hence the causal nature of the quantity $1/\varepsilon$ and, as a consequence, the dispersion relations for it. On the other hand, if we write $\vec{D} = \varepsilon\vec{E}$, one cannot consider $\vec{E}$ to be the 'cause' as it is determined by the distribution of the charges in the system (metal) itself. As a result the dispersion relations may not be satisfied for ε. It is true that for sufficiently long wavelengths ($q \to 0$; to be more precise, we are dealing with values $q \lesssim 1/L$, where L is the size of the system) they are satisfied and inequality (6) is valid (to be more precise, we are talking about the condition $\varepsilon(0,q \to 0) \geqslant 1$). However, in superconductivity theory large values of q, which reach $q_{\mathbf{F}}$ — the wavevector at the Fermi surface ($q_{\mathbf{F}} \gtrsim 10^7$ to $10^8\ \mathrm{cm}^{-1}$) — are the important ones.

It thus follows from (9) that for the values of q which are of interest to us the static ($\omega = 0$) longitudinal dielectric permittivity $\varepsilon(0,q)$ may be negative, and because of that (see (5)) values $\lambda > \mu$ are totally admissible. Moreover, in order to reach large $T_{\mathbf{c}}$ it is advantageous, and if we use the approximation (5) even necessary,* to produce materials with $\varepsilon(0,q \to q_{\mathbf{F}}) < 0$. It turned out that systems with a negative permittivity $\varepsilon(0,q)$ are not at all somewhat exotic. On the contrary, such structures as metallic hydrogen or aluminium possess (at least, in the direction of maximum symmetry) a negative value of the static longitudinal dielectric

---

* In a more general approach (when we take into account Umklapp processes and so on) very considerable values of T_{c} can probably be reached also when $\varepsilon(0,q) > 0$.

permittivity in a wide range of wavevectors.[7] The possibility of the occurrence of negative values of the static dielectric permittivity $\varepsilon(0,q)$ becomes particular clear (one can say, obvious), if we remember the difference between the average macroscopic electrical field $\vec{E}$ and the effectivee (or acting) field $\vec{E}_{\mathbf{eff}}$. In an isotropic medium $\vec{E}_{\mathbf{eff}} = \vec{E} + a\vec{P}$, where $\vec{P}$ is the polarization of the medium; in some particular cases, corresponding to the well-known Clausius-Mosotti or Lorenz-Lorentz formulae, $a = 4\pi/3$. By definition $\vec{P} = (\varepsilon - 1)\vec{E}/4\pi$, and for a medium consisting of independent 'soft' dipoles, $\vec{P} = \alpha n\vec{E}_{\mathbf{eff}}$, where α is the polarizability of the dipoles (molecules) and n their density. It is clear that under such conditions

$$\varepsilon = 1 + \frac{4\pi\alpha n}{1 - a\alpha n} , \tag{10}$$

and the value $\varepsilon < 0$ is reached when $\alpha n > 1/a$ (we assume that $a < 4\pi$). Of course, formula (10) which refers directly only to relatively small values of q (quasi-uniform field)* is given here only as an illustration, while the problem of the effective field in a metal is rather complicated (we remember, for instance, that in a dilute plasma to a good approximation $\vec{E}_{\mathbf{eff}} = \vec{E}$, that is, $a = 0$). However, the fact that in a condensed system and, in particular, in a metal a value $\varepsilon(0,q) < 0$ is admissible, can be considered to be plausible already on the basis of what we have said. When we take the difference between $\vec{E}$ and $\vec{E}_{\mathbf{eff}}$ into account (and this is therefore, in general, necessary) the theory of superconductivity becomes more complicated and has so far not been developed sufficiently (it is no longer possible to do this now on the basis of formula (5)[7]). The search for conditions and systems for which $\varepsilon(0,q) < 0$ may possibly become most important in the field of high-temperature superconductivity.

One can thus say that the problem of high-temperature superconductivity is in some sense in a transitional stage. Simple estimates and intuitive consideration, which were given in sections 2 and 3, are, I submit, justified, useful, and they give some indications as to possible directions of search. However, there is as yet no theory which is adequate for the problem. Until it is produced — and this problem is very difficult — it is impossible to be confident that there will be fast progress. Rapid success is, however, not ruled out as the search for high-temperature superconductors, albeit without

---

* It follows from theory and has been mentioned that in an equilibrium system $\varepsilon(0,q \to 0) \geqslant 1$. This is not in contradiction to what we have said, as a formula such as (10) can refer also to values of q for which a negative value of $\varepsilon(0,q)$ is already admissible.

a completely clear sense of direction, is carried out in a number of laboratories.

V. CONCLUDING REMARKS

It is characteristic and natural for physicists to tend to study first of all the simpler objects – light atoms, diatomic molecules, the simplest solids and liquids, rather than heavy atoms, polyatomic molecules, liquid crystals, polymers, multi-component alloys, or solids with a complex structure. However, the simple situations become gradually to some extent clear and – and this is not less important – for an understanding of complex systems (say, gigantic protein molecules) the knowledge of the structure of atoms is completely insufficient. It is therefore understandable that both in atomic and in molecular solid state physics there is now (and, strictly speaking has been for a number of years) going on a process of changing to a study of more and more complicated objects. This is also true for the study of superconductors. The simplest, alkali metals are under normal conditions not superconducting at all and therefore nobody studied their superconductivity. For all other metallic elements with more complex structures the critical temperature did not exceed 10 K. In the search for superconductors with higher critical temperatures and other parameters (such as, for instance, critical field) efforts turned to various alloys which even not very long ago (in particular, in my memory) were considered to be 'dirty' objects for the study of such a 'pure' phenomenon as superconductivity. Now, however, we are considering not only very new and special alloys and compounds, which are obtained under pressure and as the result of special processing, but one has gone over to yet more complicated or, at any rate, unusual systems: various organic and inorganic filamentary and layer compounds, artificial 'sandwich' type systems, some special kinds of semiconductors and semimetals, and non-equilibrium systems (without mentioning metallic hydrogen). Even biological structures have attracted attention in the field of the search for high-temperature superconductors.

In Table 2.3 opposite, we give a list – of course, in a rather arbitrary form – of the various trends, which we have mentioned, for the search for high-temperature superconductors.

The study of all these materials and systems has, in fact, only just started. Estimates of the critical temperatures for such systems are connected with additional difficulties and are, on the whole, far from reliable. Strictly speaking, as we emphasized in section 4, superconductivity theory is for the case of high values of T_c as yet in an insufficiently well developed state. It is, however, more or less clear that for all these systems and substances, at least for the equilibrium ones, there are no realistic grounds for expecting to reach values $T_c \gtrsim 1000$ K. On the other hand, $T_c \lesssim 100$ to 300 K does, in general, not meet with any objections, as far as principles are concerned. I feel that it would be especially 'unlucky' if one would not be able in at least one case to reach critical temperatures $T_c \sim 100$ to 200 K and thereby produce a high-temperature superconductor. It is a different matter whether this achievement would have practical importance and, in general, it is unclear what will be the development of the physics and material science of the new superconductors.*

TABLE 2.3

Material or system	Superconductivity mechanism (nature of the attraction between the conduction electrons)	Maximum possible value of $T_{c,max}$ (rough estimates in °K)
Metallic hydrogen or deuterium, alloys based on them	phonon	100 to 200
Organic metals, hydrogen-containing metals, and so on	phonon	50 to 100 (?)
Materials with electronic structural phase transitions	phonon or exciton	100 to 300 (?)
Three-dimensional metals with exciton bands, and so on	exciton	100 to 300 (?)
Layer compounds and dielectric-metal-dielectric sandwiches	exciton	100 to 300 (?)
Filamentary (quasi-one-dimensional) compounds	exciton	100 to 300 (?)
CuCl and CdS with impurities, obtained under pressure and as the result of special processing	?	300 (?)
Non-equilibrium superconductors		There are no clear limits, but in practice it is probable that $T_{c,max} \lesssim 1000$ K.

* Even in the helium temperature range where we known a large number of superconducting materials with different properties, and where also a large store of data has been accumulated, it is far from easy to solve the problem about the perspectives for the use of superconductivity (taking economic factors into account; see, for instance, Ref. 2).

One should, strictly speaking, treat this last statement in the sense that the problem of high-temperature superconductivity in its present stage is a physical rather than a technical or technological, problem. Moreover, in contrast, say, to the problem of controlled thermonuclear fusion, we are not even totally confident that the basic problem can in general be solved, that is, that it is possible to produce equilibrium metallic superconductors for which, say, the critical temperature $T_c \sim 100$ K. Under such conditions it is impossible to establish any dates for the 'incorporation' into the technique, and so on. However, this does not diminish interest in the problem as such and should not prevent the acknowledgement that potentially it could be exceptionally important for technical purposes. When we talk about physics, the problem of high-temperature superconductivity has in that respect almost the same rank as the problem of controlled thermonuclear fusion.

REFERENCES

1. O. Fischer, M. Derous, S. Roth, R. Chevrel, and M. Sergent. J. Phys. C, **8** (1975) L 474; S. Foner, J. Phys. Soc. Japan, **50** (1981) 2595.
2. R.A. Hein, Science, **185** (1974) 211.
3. P. Komarek, Contemp. Phys., **17** (1976) 355.
4. L.N. Bulaevskii, V.L. Ginzburg, D.I. Khomskii, D.A. Kirzhnits, Yu.V. Kopaev, E.G. Maksimov, and G.F. Zharkov, *The problem of High-Temperature Superconductivity*, (Eds. V.L. Ginzburg and D.A. Kirzhnits) Moscow, Nauka, 1977. An English translation was in 1982 published by Plenum Press, New York.
5. V.L. Ginzburg, Ann. Rev. Mat. Sc., **2** (1972) 663.
6. V.L. Ginzburg, Sov. Phys. Uspekhi, **19** (1976) 174.
7. O.V. Dolgov, D.A. Kirzhnits, and E.G. Maksimov, Rev. Mod. Phys., **53** (1981) 81; O.V. Dolgov and E.G. Maksimov, Usp. Fiz. Nauk, **135** (1981) 441; **138** (1982) 85; Sov. Phys. Uspekhi, **24** (1981) No. 10; **25** (1982) No. 9.
8. K. Anders, *et al.*, Phys. Rev. Lett., **45** (1980) 1449, 1589.
9. V.L. Ginzburg and D.A. Kirzhnits, Sov. Phys. JETP **19** (1964) 269; V.L. Ginzburg, Phys. Lett. **13** (1964) 101.
10. V.L. Ginzburg, Solid State Com., **39** (1981) 991.
11. C.B. Homan, K. Laojindapun, and R.C. MacCrone, Physica, **107** B (1981) 9.
12. B.A. Volkov, Yu.V. Kopaev, *et al.*, JETP Lett., **27** (1978) 7; **30** (1979) 285; Zh. Eksp. Teor. Fiz., **81** (1981) 729, 1904; **82** (1982) 1354; Sov. Phys. JETP, **54** (1981) 341, 1008; **55** (1982) No.4; B.A. Volkov, V.L. Ginzburg, and Yu.V. Kopaev, JETP Lett., **27** (1978) 206.

CHAPTER 3

Cosmic Ray Astrophysics

I. INTRODUCTION

Cosmic ray astrophysics has appeared only in our century and, in fact, only in the second half of this century. We are talking here of the explanation and incorporation of the rôle which is in astronomy played by cosmic rays and by the radio-, optical, X-ray and gamma-ray emission, as well as the high-energy neutrinos, which are produced by them. It is perfectly clear that the revolution which is taking place in astronomy — its change from an optical to an all-wavelength astronomy — is closely connected with the development of cosmic ray astrophysics. It is just this change together with the discovery of the non-stationary nature (expansion) of the Metagalaxy which are the most prominent events in twentieth century astronomy.

Nowadays we count as cosmic rays only charged particles with a sufficiently high (mainly relativistic) energies; this is the reason why it is impossible to reckon gamma-ray and X-ray astronomy, and also high-energy neutrino astronomy (and certainly not neutrino astronomy in general) to be part of cosmic ray astropnysics. In this connection one also uses the term high-energy astrophysics which combines all the above-mentioned branches of astronomy. However, the present lecture is devoted, mainly, to cosmic rays and we shall be interested in the electromagnetic and especially neutrino emission only in those cases where they are produced by cosmic rays. As a rule we shall therefore use the term cosmic ray astrophysics.

The term 'origin of cosmic rays' which one often encounters, especially in the English literature, will also be used, but in its narrower sense, when we are dealing with the provenance of the cosmic rays which are observed on the Earth.

We do not include in our programme any detailed exposition of the history of the study of cosmic rays, either as a whole, or as far as astrophysical problems are concerned (see Refs. 1 to 3). Nonetheless it is appropriate to start with a brief historical survey.

The study of cosmic rays started, in fact, around 1900 as the result of observations of background ionization in a gas in sealed vessels: the question arose whether this ionization could be completely explained by the radioactive emission from the Earth's surface, the walls of the vessel, and the radioactive emanations in the gas. It was not an easy problem to solve, especially as far as the explanation of the rôle of the Earth was concerned. To find out about this manned balloon flights were undertaken. They led to the undoubted discovery of cosmic rays by V. Hess in 1912. Particularly successful was a flight by Hess on 7th August 1912 when he reached a height of about 5 km; the speed of the ionization increased in that case by a considerable factor as compared with that which was observed at sea level. These results were confirmed in 1913 – 1914 by W. Kolhörster who reached a height of 9 km where the speed of the ionization was still much higher. However, when we mention the 'undoubted discovery' we have in mind the establishment of the facts and, to some extent, contemporary ideas about the structure of the atmosphere. At that time it was assumed as one of the possibilities, for instance, that the increase in the ionization speed when one got away from the Earth's surface was connected with the presence in the upper atmosphere of an appreciable amount of radioactive emanations and not with the action of some unknown cosmic radiation. Somehow or other it was not before around 1927 that all doubts were removed about the existence of cosmic rays – a penetrating 'radiation' of extraterrestrial origin – as the result of the demonstration of the fact that these 'rays' were absorbed appreciably more weakly than the gamma-rays from radioactive elements. However, even if there were for, at least, 15 years doubts about the existence of cosmic rays, their nature at that time was considered to be clear – they 'must be' hard gamma-rays (this conclusion was connected with the fact that it was just the gamma-emission from radioactive elements which was the most penetrating emission). However, in 1927 the geomagnetic effect – that is the geomagnetic latitude dependence of the ionization produced by the cosmic rays – was discovered. As a result it became finally clear around 1936 that the primary cosmic rays are charged particles. Initially they were thought to be electrons, but afterwards it became clear, in 1939 – 1941, that protons play the main rôle. Finally, in 1948 it was possible to ascertain that in the composition of the primary cosmic rays there are also nuclei of a whole range of elements.

It has thus been necessary for around 40 years to discover, even if only in very rough outlines, what the primary cosmic rays are. We have stressed this, in particular, with the aim of emphasizing that many scientific problems and questions need for their solution several decades. Of course, such a conclusion is no news. Nonetheless, the majority of physicists and astronomers, because of their youth, consider discoveries which are twenty, and even more those which are thirty, years old often to belong to the archaeological antiquity. Sometimes events, indeed, develop very swiftly. But one encounters also a different situation. In one of the preceding lectures I have already discussed this question and, in particular, dwelled upon a few examples (the discoveries of pulsars and quasars). I shall not again develop this theme.

As I said, up to 1950 the composition of the primary cosmic rays was known in rough outline. Several papers appeared also which foresaw the potential importance of cosmic rays for astrophysics. For instance, Baade and Zwicky[4] connected in 1934 the occurrence of supernovae with the formation of neutron stars and the generation of cosmic rays. In 1949 Fermi[5] approached cosmic rays as a gas of relativistic particles moving in interstellar fields. Nonetheless, on the whole, the rôle of cosmic rays in astronomy remained altogether obscure and, in practice, only physicists were interested in cosmic rays. As the main reason for this one can see the high degree of isotropy of the cosmic rays (we assume the effect of the Earth's magnetic field to be eliminated). For this reason even the detailed conclusions themselves about the composition and energy spectrum of the cosmic rays at the Earth told us little about their sources and especially about the localization of these sources. The situation is here analogous to the one which would occur, if we knew only the spectrum of all stars taken together, while the separate stars were not observed.

The birth of cosmic ray astrophysics and of high-energy astrophysics on the whole can therefore be assigned only to 1950 – 53 when the picture changed radically. It was then that the synchrotron nature of a considerable fraction of the cosmic radio-emission became clear. As a result it became possible to obtain extensive information about the electron component of the cosmic rays far from the Earth – in our Galaxy and beyond its limits. Moreover, at the price of a few assumptions one could use the intensity of the synchrotron radiation to estimate the total energy of the cosmic rays in the sources (in supernova shells, in radio-galaxies, and so on; *vide infra*). An understanding of these points took, if we consider the astronomical community as a whole, around ten years and, at any rate, at the Paris symposium on radio-astronomy in 1958[6] the close connection between radio-astronomy and cosmic rays was no longer disputed. The preceding history which is rather interesting and is often presented completely incorrectly as a result of unfamiliarity with the original literature was briefly discussed in the introduction to Ref. 7 and in more detail in a paper by me which is included in the collection of Ref. 3.

II. COSMIC RAYS AND COSMIC ELECTROMAGNETIC RADIATION

It is now necessary to discuss, albeit briefly, the connection between cosmic rays and cosmic radio-emission (for details, see, for instance, Refs. 7 to 9).

In a constant, uniform magnetic field of strength $\vec{H}$ a particle of charge e and mass m moves in a spiral with rotational frequency equal to

$$\omega_{\mathrm{H}}^{*} = \omega_{\mathrm{H}} \frac{mc^2}{E} = \frac{|e|H}{mc} \frac{mc^2}{E} = 1.76 \times 10^7 \, H \frac{mc^2}{E} \, \mathrm{s}^{-1} , \tag{1}$$

where E is the total energy of the particle and where in writing down the numerical factor we assumed the particle to be here, and henceforth, an electron, while the field H is measured in oersted (in the conditions which are of interest to us the field $\vec{H}$ is the same as the magnetic induction $\vec{B}$). In the ultra-relativistic case, that is, when $E \gg mc^2$, such a particle at not too small an angle to the field (it is necessary that the angle χ between the field $\vec{H}$ and the particle velocity $\vec{v}$ satisfies the condition $\chi \gg mc^2/E$) emits electromagnetic waves with many frequencies ω which are overtones of the frequency $\omega_{\mathrm{H}}^{*}/\sin^2\chi$; in practice we can assume the spectrum of the radiation to be continuous. The maximum in the radiation spectrum of a single electron (to fix the ideas we talk about an electron, although the expressions in symbols are general in nature) occurs at the frequency

$$\nu_{\mathrm{m}} = \frac{\omega_{\mathrm{m}}}{2\pi} = 0.07 \frac{|e|H_{\perp}}{mc} \left(\frac{E}{mc^2}\right)^2 = 1.2 \times 10^6 \, H_{\perp} \left(\frac{E}{mc^2}\right)^2$$

$$= 1.8 \times 10^{18} \, H_{\perp} \, (E\,(\mathrm{erg}))^2 = 4.6 \times 10^{-6} \, H_{\perp} \, (E(\mathrm{eV}))^2 \; \mathrm{Hz}, \tag{2}$$

where $H_{\perp} = H \sin\chi$ is the component of the field $\vec{H}$ at right angles to the particle velocity and where we have substituted for the numerical values the values of the electron charge and mass for e and m.

At the frequency ν_{m} the spectral power density of the radiation is

$$p_{\mathrm{m}} \equiv p(\nu_{\mathrm{m}}) = 1.6 \frac{|e|^3 H_{\perp}}{mc^2} = 2.16 \times 10^{-22} \, H_{\perp} \; \mathrm{erg\,s^{-1}\,Hz^{-1}}. \tag{3}$$

Let us consider a volume in which the magnetic field is on average isotropic in direction while the radiating electrons also are distributed isotropically in velocity. The radiation will then also be isotropic while the radiative emissivity ε_ν, that is, the spectral power of the radiation per unit volume and unit solid angle, is for mono-energetic electrons equal to

$$\varepsilon_\nu = \frac{p(\nu)}{4\pi} N_e , \tag{4}$$

where $p(\nu)$ is the spectral radiation density from a single electron in all directions, and N_e is the density of relativistic electrons. It is clear from (3) and (4) that the maximum emissivity is equal to

$$\varepsilon_{\nu,m} = \frac{p_m}{4\pi} N_e = 0.13 \frac{|e|^3 H_\perp}{mc^2} N_e = 1.7\times 10^{-23} H_\perp N_e \frac{\text{erg}}{\text{cm}^3\text{s sterad Hz}} , \tag{5}$$

where $H_\perp$ is some average value of $H_\perp$ for the emitting region.

The maximum radiation intensity along the line of sight which we obtain from isotropically distributed mono-energetic electrons is equal to

$$I_{\nu,m} = \int \varepsilon_{\nu,m} \, dR = 1.7\times 10^{-23} H_\perp \tilde{N}_e \frac{\text{erg}}{\text{cm}^2\text{s sterad Hz}}$$

$$= 1.7\times 10^{-26} H_\perp \tilde{N}_e \frac{\text{W}}{\text{m}^2 \text{ sterad Hz}} , \tag{6}$$

while

$$\tilde{N}_e = \int N_e(\tilde{R}) \, dR = N_e L \tag{7}$$

is the total number of emitting electrons along the line of sight; we have here, clearly, assumed that the component of $H_\perp$ along the line of sight is constant, or we understand by $H_\perp$ an appropriate average. When we changed to the last expression in (7) we assumed that the electrons with density N_e occupy a region of size L along the line of sight.

Formula (6) is convenient for estimating, for known I_ν and $H_\perp$, the minimum necessary values of $\tilde{N}_e$ and N_e. The fact that we are dealing just with the minimum values is caused by the use of the maximum power $\varepsilon_{\nu,m}$.

Under cosmic conditions we can meet only exceptionally with mono-energetic particles; as a rule we are dealing with an energy distribution. Very often we encounter and use power-law spectra. To be concrete, one can often, in applications to relativistic electrons which are isotropic as far as velocity directions are concerned, approximate the spectrum by the expression

$$N_e(E)\, dE = K_e E^{-\gamma_e} \, dE \quad , \quad E_1 < E < E_2 . \tag{8}$$

For such a spectrum and for the case of a magnetic field which on average is isotropic along the line of sight the radiation intensity is equal to $(\gamma \equiv \gamma_e)$

$$I_\nu = a(\gamma)\,\frac{|e|^3}{mc^2}\left(\frac{3|e|}{4\pi m^3 c^5}\right)^{\frac{1}{2}(\gamma-1)} L K_e H^{\frac{1}{2}(\gamma+1)}\,\nu^{-\frac{1}{2}(\gamma-1)}$$

$$= 1.35\times 10^{-22}\,a(\gamma)\,L K_e H^{\frac{1}{2}(\gamma+1)}\left(\frac{6.26\times 10^{18}}{\nu}\right)^{\frac{1}{2}(\gamma-1)}\ \frac{\text{erg}}{\text{cm}^2\,\text{s sterad Hz}}\,, \qquad (9)$$

where K_e is the coefficient in (8), L the size of the emitting region along the line of sight (in fact, the values K_e and $H^{\frac{1}{2}(\gamma+1)}$ are appropriate averages) and $a(\gamma)$ is a function given in Refs. 7 to 9 (here it is sufficient to mention that when $\gamma \approx 1.5$ to 5 the value $a(\gamma) \sim 0.1$).

It is clear from (9), first of all, that for power-law electron spectra the radiation spectrum is also a power law, and

$$I_\nu \sim \nu^{-\alpha}\ , \quad \alpha = \tfrac{1}{2}(\gamma-1)\ , \qquad (10)$$

that is, α depends solely on γ.

Measuring the value of the intensity I_ν itself one can, as in the case of a mono-energetic spectrum, find $K_e L$, if one knows the field $\vec{H}$, and thus also the number of emitting electrons in the corresponding energy range.

In the foregoing we had in mind emission which had an angular distribution, but one can also easily obtain formulae for discrete sources[7-9]. Here only one thing is important for us: from the data about the cosmic radio-emission (in the majority of cases we are interested just in the radio-band)* we get information about relativistic electrons far from the Earth – in the Galaxy, in supernova shells, in radio-galaxies, and so on. If we forget the fact that one must average along the line of sight and that it is necessary somehow or other to know the size of the emitting region, to infer the density of the relativistic electrons we must still give the magnetic field strength in the emitting region. As far as the field direction is concerned we can get some information from polarization measurements of the same synchrotron radiation. However, it is necessary to determine the field strength itself independently. The determination or estimate of the strength of the cosmic

---

* To elucidate the problem of the frequency band in which the emission mainly takes place in a particular case it is usually sufficient to use formula (2). For instance, for the Galaxy as a whole values of $H \sim 10^{-6}$ to 10^{-5} Oe are characteristic, while in the composition of the electron component of the cosmic rays there are certainly electrons with energies $E_e \sim 10^9$ to 10^{12} eV = 1 to 1000 GeV. When $H = 3\times 10^{-6}$ Oe and $E = 10^9$ to 10^{12} eV, the frequency is, according to (2), $\nu_m \sim 10^7$ to 10^{13} Hz and the wavelength $\lambda_m = c/\nu_m$ lies in the band from 30 m to 3×10^{-3} cm = 30 μm.

magnetic fields for various objects can be made by different methods. This has to be the case, if we bear in mind that the fields one encounters themselves vary greatly — for pulsars the value of H reaches 10^{12} to 10^{13} Oe, for the Sun and a number of stars $H \sim 1$ to 10^4 Oe, in the galactic disk $H \sim 10^{-5}$ Oe, in the halo the field is probably somewhat less, and in the intergalactic space, $H \leqslant 10^{-7}$ to 10^{-8} Oe. A rather general principle which enables us to estimate the field H under quasi-stationary conditions and, concretely, in the Galaxy and in supernova shells is the assumption about approximate equality of the energy densities of the cosmic rays and of the magnetic field

$$w_{\mathbf{c.r.}} \sim w_{\mathbf{H}} \sim \frac{H^2}{8\pi} . \tag{11}$$

There are several data which give indications favourable to the validity of this relation (for instance, at least in some regions of the Galaxy, $w_{\mathbf{c.r.}} \sim 10^{-12}$ erg/cm^3 while from independent data we find $H \sim 3$ to $5\,\mu$Oe, that is, also $w_{\mathbf{H}} = H^2/8\pi \sim 10^{-12}$ erg/cm^3), and one also has the following arguments. Cosmic rays are 'frozen in' in the highly conducting interstellar gas. Therefore, if $w_{\mathbf{c.r.}} \gg H^2/8\pi$ (and, hence, the pressure of the cosmic rays $p = w_{\mathbf{c.r.}}/3 \gg H^2/8\pi$) cosmic rays cannot be confined by the field and they will begin to 'leak out' from the region with the weaker field. In the opposite case, when $H^2/8\pi \gg w_{\mathbf{c.r.}}$ one cannot, in general, guarantee quasi-stationarity — under the action of the sources $w_{\mathbf{c.r.}}$ will increase with time.

Somewhat generalizing relation (11) we can assume that

$$\kappa_{\mathbf{H}} w_{\mathbf{c.r.}} = w_{\mathbf{H}} = H^2/8\pi , \tag{12}$$

where $\kappa_{\mathbf{H}}$ is a coefficient which is characteristic for the region or object under consideration; it is clear that in the case (11) $\kappa_{\mathbf{H}} \sim 1$. We note that we can use, and sometimes more reasonably, the total energy in the source (for instance, in a supernova shell) instead of the energy density, that is, we can put

$$\kappa W_{\mathbf{c.r.}} = W_{\mathbf{H}} , \quad W_{\mathbf{c.r}} = w_{\mathbf{c.r}} V , \quad W_{\mathbf{H}} = \frac{H^2}{8\pi} V , \tag{13}$$

where $w_{\mathbf{c.r.}}$ and $H^2/8\pi$ are the average values in a source with volume V (one can, of course, also refrain from introducing averages and write $W_{\mathbf{c.r.}}$ in the form $W_{\mathbf{c.r.}} = \int w_{\mathbf{c.r.}} dV$, and so on). Unfortunately, assumption (11) is still insufficient as only the electrons are responsible for the synchrotron radio-emission (apart from special cases). It is known that at the Earth the energy density of the electron component of the cosmic rays

$$w_{\mathbf{c.r.,e}} \sim \frac{1}{100} w_{\mathbf{c.r.}} . \tag{14}$$

In the more general case we can put

$$\kappa_e w_{c.r.,e} = w_{c.r.} \quad (15)$$

If we assume that the coefficients κ_H and κ_e are known, then, clearly, we can find from the cosmic radio-emission data for both $w_{c.r.}$ and $H^2/8\pi$ (or $W_{c.r.}$ and $W_H = \int (H^2/8\pi)\, dV$). For example, it is usual to proceed, putting $\kappa_H \sim 1$ and $\kappa_e \sim 100$ (see (11) to (15)). Undoubtedly, such an approach has only limited value and the problem of determining H or the density $w_{c.r.}$ (or, in other words, determining the coefficients κ_H and κ_e) independently is a topical problem. Apart from the ones already mentioned there are some other possibilities in this direction and we shall be dealing with them below.

It is now appropriate to mention what basic conclusions followed, already more than 20 years ago, from the establishment of a connection between radio-astronomy and cosmic rays.

Firstly, it became clear that the generation of cosmic rays is a universal phenomenon — cosmic rays exist in interstellar space and in supernova shells and in other galaxies, especially in radio-galaxies. Hence it follows also that cosmic rays are valuable sources of astronomical information, not only directly (I have in view the cosmic rays at the Earth) but also as the result of the possibility of detecting the radiation produced by the cosmic rays. Initially one was dealing basically with radio-emission. However, subsequently optical, X-ray, and gamma-ray emission, as well as, in principle, high-energy neutrino emission were added to this.

Secondly, it became clear how important cosmic rays are as far as energetics and dynamics are concerned. For instance, their energy density in the Galaxy, $w_{c.r.} \sim 10^{-12}$ erg/cm^3 is, as we have already mentioned, of the same order of magnitude as the energy density of the interstellar field $w_H = H^2/8\pi$ and the energy density of the interstellar gas $w_g = \frac{3}{2} nkT$ (for example the density $w_g \sim 10^{-12}$ erg/cm^3 for a gas density $n \sim 1$ cm^{-3} and a temperature $T \sim 10^4$ K). In a number of objects the density $w_{c.r.}$ exceeds, or may exceed, the densities w_H and w_g. The same also holds for the cosmic ray pressure $p_{c.r.} = w_{c.r.}/3$ (relativistic particles dominate in cosmic rays, which are practically isotropic). An estimate of the total energy of the cosmic rays in our Galaxy $W_{c.r.} \sim 10^{56}$ erg, while in powerful radio-galaxies

$$W_{c.r.} \leqslant 10^{61} \text{ erg} \sim 10^7 M_\odot c^2$$

(*vide infra*).

Both conclusions given here are completely in accordance with contemporary ideas from plasma physics: in a dilute plasma in which there are particle beams, shock waves, and various magnetic inhomogeneities one should just expect an efficient acceleration of some fraction of the particles and also their scattering and diffusion. It is difficult to overestimate the value of these facts for astronomy. As cosmic rays are such an important ingredient in the cosmos the rôle of high-energy astrophysics as a whole is obvious. This is also true of the somewhat narrower field of studies — cosmic ray astrophysics. Simultaneously, as usually is the case in such situations, there is an interpenetration of the various fields and trends and to say at this moment where high-energy astrophysics and cosmic ray astrophysics begin and where they end is not very easy. Anyhow, why should we make such a distinction at all ?

In the last ten years the most important facet of high-energy astrophysics has been the development and, indeed, the appearance of gamma-astronomy observations.* From the point of view of the study of cosmic rays the detection of gamma rays from the decay of π^0-mesons and, to some extent, of other unstable particles, is of particular value. They are produced when the proton-nuclear component of the cosmic rays collides with the nuclei in the gas. The intensity I_{γ,π^0} of these gamma rays is proportional to the gas density n and to the cosmic ray intensity $I_{\mathrm{c.r.}}$ or, after some calculations, to their energy density $w_{\mathrm{c.r.}}$. Thus the measurement of I_{γ,π^0} opens up, indeed, the only known possibility of directly determining the density $w_{\mathrm{c.r.}}$ far from the Earth. The analogy with the registration of synchrotron radiation which enables us to determine the energy density $w_{\mathrm{c.r.,e}}$ (for a known field H) is here completely appropriate. We shall mention later, in section 3, one important result which was obtained along such a gamma-astronomy road. It is impossible not to mention also other possibilities of gamma-astronomy, for instance, the observation of nuclear gamma-lines and of annihilation radiation (the line $E_\gamma = 0.51$ MeV), the registration of gamma-radiation formed in bremsstrahlung or the inverse Compton effect for electrons (for instance, in quasars), ground-based observations of gamma rays with $E_\gamma \gtrsim 10^{11}$ to 10^{12} eV (from Cherenkov emission flashes in the atmospheres). In general, one can say that the 'gamma-window' into the cosmos has been widely opened, taking into account the experimental possibilities which already exist and, undoubtedly, will be more and more widely used in astronomy.

From this point of view of obtaining information about cosmic rays, or, more precisely, about their electron component using X-ray astronomy methods,

---

* The progress of X-ray astronomy has been even more spectracular, but this field in some sense is 'atypical' for high-energy astrophysics. I mean that the main part of cosmic X-rays come from a hot, but non-relativistic plasma ($T \leqslant 10^9\,\mathrm{K} \sim 10^5$ eV) and is not directly connected with cosmic rays.

of particular interest is the X-ray emission which is produced when relativistic electrons are scattered by the thermal relict radiation with a temperature $T \approx 3\,K$. This process (usually called the inverse Compton effect) is in many respects analogous to synchrotron radiation but now the rôle of the magnetic field energy density $H^2/8\pi$ is played by the energy density of the thermal (black-body) radiation $w_{ph,T}$. As the energy density of the relict radiation is known (for a temperature of 3 K this energy density $w_{ph,T} = 6.1 \times 10^{-13}$ erg/cm^3) measurements of the intensity of the scattered X-ray radiation allow us to determine the intensity and energy density $w_{c.r.,e}$ of the relativistic electrons. If the same relativistic electrons from the same region of space produce measurable synchrotron radiation, we can also determine the magnetic field strength in the region studied. Unfortunately, this method has so far hardly been used. Sometimes (for example, for the Crab nebula) synchrotron radiation also falls in the X-ray region.

AAs far as gamma-radiation is concerned, we have already mentioned the component with intensity I_{γ,π^0} which is generated by the proton-nuclear component of the cosmic rays and is mainly formed as the result of the decay of π^0 mesons. Moreover, there is a gamma-component which is the bremmstrahlung of the relativistic electrons in the interstellar medium. The energy of the corresponding gamma-photons E_γ is then of the order of (but, of course, not larger than) the energy of the electrons producing them. Therefore, in the main part of the accessible gamma-band ($E_\gamma \leqslant 100$ to 200 MeV) we may get information about electrons with energies $E_e \leqslant 200$ MeV which make only a small contribution to the observed galactic radio-emission.

The study of cosmic rays by gamma- and X-ray-astronomy methods is, essentially, in the initial phase. At the same time there is no doubt that just along this road we can in many respects supplement and refine radio-astronomical data. High-energy cosmic neutrinos are generated only by cosmic rays, in the main by their proton-nuclear component. Apart from theoretical ones, there are as yet no results in this field, but the perspectives are very impressive.

III. ORIGIN OF THE MAIN PART OF THE COSMIC RAYS OBSERVED AT THE EARTH

Cosmic rays which reach the solar system are clearly distinguished in that we can observe them directly. At the present this is done by applications of balloons which enable us to study primary cosmic rays, practically beyond the limits of the atmosphere and with rocket measurements and measurements from satellites and interplanetary probes which are reaching even the periphery of the solar system. All those cosmic rays we will give the same name, for brevity, namely, we shall call them the primary cosmic rays observed at the Earth. In the composition of the primary cosmic rays there is a component which comes from the Sun, but for energies $E \geqslant 1$ GeV their share is small. The main part of the primary cosmic rays therefore come to us from interstellar space. There are weighty grounds for assuming that all those cosmic rays are formed in our Galaxy except, maybe, particles with very high energies ($E > 10^{17}$ to 10^{19} eV). In what follows I shall hardly at all touch upon the problems of cosmic rays of solar origin or upon the modulational effects which occur in the solar system. Cosmic rays with very high energies are, for instance, considered in Ref. 10. In view of the fact that the cosmic ray spectrum is decreasing the contribution from cosmic rays with very high (superhigh) energies to the flux (or the intensity) and to the energy density of all cosmic rays is completely negligible. A similar contribution from particles of solar origin is somewhat larger, but also insignificant. We can therefore talk about the main part of the cosmic rays, observed at the Earth, having in mind in that case all particles except the solar and very-high-energy ones.

One of the central problems in cosmic ray astrophysics is that of their origin, and in what follows we shall be thinking, without further stipulations, of the origin of the main part of the cosmic rays, observed at the Earth.

To give an answer to this question means, firstly, to indicate, even if only roughly, the position of the cosmic ray sources, for instance, by stating that they are in the galactic disk. Secondly, we must, of course, identify the sources, for instance, by connecting them with supernova shells. We must, however, emphasize that a knowledge of the nature of the sources is not all that important from the point of view of being able to explain a whole range of other problems. For instance, it has no particular value for solving the third problem, which is, to establish the 'confinement region' of the cosmic rays in our Galaxy. Furthermore, there are also many other problems: how do the cosmic rays move (propagate) from the sources to the Earth, what transformations of their chemical and isotopic composition takes place, what is the rôle of plasma and magneto-hydrodynamic effects on the propagation of the cosmic rays, and so on and so forth. One must answer all these questions using the existing general astrophysics information, on the basis of radio- and gamma-astronomical data (see section 2 above) and using our knowledge about the primary cosmic rays at the Earth.

The last kind of information reduces first and formost to a knowledge of the intensity $I_{Z,A}(E)$ for nuclei of atomic number Z and mass number A

(in practice, we must usually be satisfied with knowing the intensity $I_Z(E)$ for all nuclei of given Z without a division into isotopes, or even the intensity $I_{c.r.}(E)$ for all cosmic rays). The first results have already appeared for the antiproton intensity $I_{\bar{p}}(E)$. Finally, the electron-positron component intensity $I_e(E) = I_{e-} + I_{e+}$ is measured with increasing accuracy. Although the positron flux is only a fraction (perhaps, up to ten per cent) of the electron flux, separate determinations of the positron intensity $I_{e+}(E)$ or of the ratio $I_{e+}/(I_{e-} + I_{e+}) = I_{e+}/I_e$ is of great interest.

We have here assumed the intensity to depend only on the total energy E of the relevant particles (one can, of course, also use the kinetic energy E_k and sometimes the energy per nucleon ε or ε_k). We can proceed in this way to a first (and very good) approximation because of the high isotropy of the cosmic rays. For this reason it is sufficient to introduce apart from the intensity $I(E)$ the degree of anisotropy $\delta(E)$ to characterize the primary cosmic rays. The relevant information about $I(E)$ and $\delta(E)$ is summarized, for instance, in Refs. 1, 8, 9, or 11. New data are rather extensively given in the proceedings of recent conferences and symposia.[12,13]

The problem of the origin of the cosmic rays arose, strictly speaking, immediately after their discovery. It is natural that during the past seven decades many hypotheses have been put forward which could not withstand the comparison with experimental and observational data which appeared later. For instance, a model in which the cosmic rays originated on the Sun has been discussed[2]*. However, already in the fifties, after the appearance of radio-astronomical data about cosmic rays far from the Earth it became clear that solar models were unacceptable (for more details, see Ref. 9). The situation is more complicated with regard to metagalactic models. In such models the main sources of cosmic rays are assumed to be metagalactic and, hence, the cosmic rays observed on Earth first of all must penetrate into our Galaxy from the Metagalaxy and after that reach the solar system.

One can counter the metagalactic model, first of all, by giving objections of an energetic nature. The fact is that in metagalactic models of cosmic ray origin cosmic rays must have approximately the same characteristics outside our Galaxy, that is, in intergalactic (metagalactic) space, at least in the vicinity of our Galaxy (say, within the limits of the Local Group of galaxies or of the Local Supercluster) as within our Galaxy. In particular, they must have the same energy densities.

---

* In the literature one often finds the term 'theory of the origin of cosmic rays'. However, to speak in such cases of a 'theory' can only be done in a very relative sense and I shall use the term 'model'.

$$w_{\mathrm{c.r.,Mg}} \sim w_{\mathrm{c.r.,G}} \sim 10^{-12}\ \mathrm{erg/cm^3}\ . \tag{16}$$

The density (16) is relatively high *(vide supra)* and it is very difficult to envisage a model in which an appreciable part of the Metagalaxy is filled by cosmic rays with a density (16). Estimates lead, rather. to the conclusion that the inequality

$$w_{\mathrm{c.r.,Mg}} \ll w_{\mathrm{c.r.,G}} \sim 10^{-12}\ \mathrm{erg/cm} \tag{17}$$

is very easily satisfied.[9] To use only energy and a few other related considerations is, however, insufficient and this explains the vitality of metagalactic models. A reliable refutation of them requires direct data about the intensity of cosmic rays beyond the limits of our Galaxy or at its periphery.

The first step in that direction was successfully made after the discovery in 1965 of the relict thermal radiation ($T \approx 3$ K). Inverse Compton losses of relativistic electrons in the field of that radiation are so strong that electrons with energies $E_{\mathrm{e}} \gtrsim (1 \text{ to } 3) \times 10^{10}$ eV can not reach the solar system even from the radio-galaxy which is the closest to us, Centaurus A (at a distance $R \approx 4$ Mpc). Indeed, the energy change of an electron in the field of black-body radiation with a density $w_{\mathrm{ph,T}}$ and a magnetic field with an average value of its energy density $w_{\mathrm{H}} = H^2/8\pi$ is given by the expression:

$$\left.\begin{aligned} E_{\mathrm{e}}(t) &= \frac{E_0}{1 + E_0 \beta t}\ , \\ \beta &= \frac{32\pi e^4}{9 m^4 c^7}\left(w_{\mathrm{ph,T}} + \frac{H^2}{8\pi}\right) = 4 \times 10^{-2}\left(w_{\mathrm{ph,T}} + \frac{H^2}{8\pi}\right)\ \mathrm{erg^{-1}\ s^{-1}}\ , \end{aligned}\right\} \tag{18}$$

Therefore, independent of the energy E_0 (at $t = 0$), after a lapse of time $t = T_{\mathrm{e}}$ the electron cannot have an energy exceeding

$$E_{\mathrm{c}} = E_{\mathrm{e,max}}(T_{\mathrm{e}}) = \frac{1}{\beta T_{\mathrm{e}}} \doteq \frac{1.56 \times 10^{13}}{[w_{\mathrm{ph,T}} + (H^2/8\pi)]\,T_{\mathrm{e}}}\ \mathrm{eV}\ , \tag{19}$$

where $w_{\mathrm{ph,T}} + (H^2/8\pi)$ is measured in erg/cm^3 and T_{e} in seconds.

From this it is clear that the maximum distance which an electron with observed energy E_{e}(eV) can traverse during a time T_{e} is equal to

$$R_{\max} = cT_{\mathrm{e}} = \frac{c}{\beta E_{\mathrm{e}}} = \frac{4.7 \times 10^{23}}{[w_{\mathrm{ph,T}} + (H^2/8\pi)]\,E_{\mathrm{e}}\ (\mathrm{eV})}\ \mathrm{cm}\ . \tag{20}$$

Substituting here the value

$$w_{\mathrm{ph,T}} + (H^2/8\pi) \geqslant w_{\mathrm{ph,T}} = 6 \times 10^{-13}\ \mathrm{erg/cm^3}$$

corresponding to a radiation temperature $T = 3K$, we get

$$R_{max} \leqslant 8\times10^{35}/E_e\ (\text{eV})\ \text{cm}$$

and thus

$$R_{max} \leqslant 10^{25}\ \text{cm} \sim 4\ \text{Mpc} \quad \text{for}\ E_e \geqslant 10^{11}\ \text{eV}\ .$$

Bearing in mind that the average speed of an electron in the given direction (the direction from Centaurus A to the solar system) in its motion towards the Galaxy, and especially within the limits of the Galaxy, in all probability is appreciably less than c we see at once that from Centaurus A electrons with $E_e \gtrsim 10^{10}$ eV, and possibly also with $E_e \gtrsim 10^9$ eV can hardly reach the solar system.

There is therefore no doubt that the electron component of the cosmic rays observed at the Earth, at least for not too low energies, must be of galactic origin. From a number of considerations which we shall not discuss here (see Ref. 14) it has already in this connection become very probable that the proton-nuclear component of the cosmic rays is also of galactic origin. Nonetheless, it is as yet not possible logically to assume that every model one can think up of a metagalactic origin of the proton-nuclear component has to be ruled out.[15] Only the gamma-astronomical method is adequate to reach such a totally conclusive refutation

In the metagalactic models the relation (16) must be satisfied everywhere and, in particular, in the Magellanic Clouds. Under such conditions one can rather reliably predict the flux of gamma-rays from the π^0-decay which must leave each of the Clouds.[14,16] For instance, for the Large Cloud (LMC) the flux

$$F_\gamma(E_\gamma > 100\ \text{MeV}) = 2\times10^{-7}\ \text{photons/cm}^2\ \text{s}$$

when we take into account only the atomic hydrogen in the Cloud. If we take into account molecular hydrogen and the gamma-rays from electron bremsstrahlung we arrive, clearly, at a larger flux. If, however, the cosmic rays are produced in the Clouds themselves, they will probably relatively fast enter interstellar space, because of the small size of these Clouds. Under such conditions one would find, for example, for the Large Cloud a flux which is smaller than the above-mentioned lower limit. The corresponding measurements are becoming possible in the next few years, but already now we can use another, related method for solving this problem. In fact, the value of the cosmic ray density in our Galaxy, $w_{c.r.,G} \sim 10^{-12}$ erg/cm^3 must in metagalactic models be constant in the whole system, including the regions at its periphery and, in particular, it should not decrease when we go from the solar system in the direction of the galactic anti-centre.[17] At the moment, the existing observations of the gamma-ray intensity $I_{\gamma}(E_{\gamma} > 100\ \text{MeV})$ in the direction of the anticentre indicate a decrease in the cosmic ray intensity (and, of course, of their energy density $w_{\mathrm{c.r.}}$) when one goes away from the solar system. It is true that so far the reliability of the results leaves something to be desired in a quantitative sense (see Ref. 13), but undoubtedly there are indications of a decrease in the density.

From the combination of all data and arguments it seems to us that metagalactic models have been refuted, although it is still desirable to substantiate such a conclusion by new gamma-astronomical measurements in the future.

In view of what we have said we believe that one needs consider only galactic models of the origin of cosmic rays (we shall not repeat reservations about the solar cosmic rays or about particles with very high energies).

IV. GALACTIC MODEL WITH A HALO

In the galactic models the cosmic ray sources are, of course, situated in our Galaxy. As far as the 'confinement region' is concerned, that is that region in which (within the limits of the Galaxy) the cosmic rays are concentrated, this is to some extent an independent problem.

For a long time already supernova outbursts have been assumed to play the rôle of the main sources. Such an assumption is fully admissible from an energy point of view. Moreover, it is known that supernova shells are powerful sources of synchrotron radiation and, hence, contain relativistic electrons. It is sufficient that in the shells (including the pulsars inside them) the proton-nuclear component is generated with a power which is one or two orders of magnitude larger than that of the electron component in order that supernovae may guarantee the quasi-stationarity of the cosmic rays in our Galaxy (*vide infra*). Nonetheless there is a hypothetical element here and we must take into account that cosmic rays to some degree can also be accelerated in interstellar space at shock wave fronts. As a result the problem of the main sources of cosmic rays in our Galaxy remains to some extent open. In our opinion supernovae remain altogether the most probable main sources, but the answer to the problem of the relative rôle of different 'channels' along which the acceleration in the supernovae takes place is less well determined. For instance, it may occur in the process of the stellar explosion itself, in the pulsar magnetosphere which remains after the explosion, in the young supernova shell and, finally, in shock waves (the fronts of old shells) which are already far from the place where the explosion took place — in interstellar space. In the past[9,18] acceleration in young shells was preferred, but now there is less unanimaty in the literature, although it is not the case that one can speak of a refutation of the earlier point of view. There has also been already for a few decades a debate about the problem of the rôle of other stars apart from supernovae (novae, A-type magnetic stars, O-type hot stars, binaries, and so on). As even such a relatively quiescent star as the Sun generates cosmic rays there is no doubt that cosmic rays will be accelerated by more active stars. However, this does not mean that such stars can guarantee the power generation which is needed for the Galaxy as a whole (see below), or accelerate cosmic rays to the highest energies for which it is not yet necessary to invoke metagalactic sources (we are thinking here about energies, at least, up to 10^{17} eV).

If the acceleration and injection of cosmic rays is due not only to supernovae the picture becomes only richer. Such a possibility is not refuted, but the assumption that supernovae play the dominating rôle, let us repeat, remains completely possible and is, in our opinion, the most plausible one.

Whichever stars, apart from the supernovae, are responsible for the acceleration of the cosmic rays, such sources are concentrated in the galactic disk and are, basically in a layer of half-thickness $h_{\mathbf{d}} \sim 100$ to 200 pc (to be more precise, the gas is concentrated in a layer with $h_{\mathbf{d}} \approx 75$ to 125 pc; the distribution of the stars depends on their type but young stars and the brightest stars are, generally speaking, situated where there is much gas). It is clear that the 'confinement region' of the cosmic rays (that is, that region where their density is large) cannot be appreciably smaller than the region occupied by the sources. In the past one has often assumed that the cosmic rays are confined to a disk with $h \sim h_{\mathbf{d}} \leqslant 200$ pc; this corresponds to the disk models of the origin of cosmic rays.

In such models the volume $V_{\mathbf{d}}$ filled by the cosmic rays and the corresponding total energy $w_{\mathbf{c.r.,d}}$ have the following estimated values

$$w_{\mathbf{c.r.,d}} \sim w_{\mathbf{c.r.}} V_{\mathbf{d}} \sim 10^{55}\ \text{erg} \quad , \quad V_{\mathbf{d}} \sim 4\pi h_{\mathbf{d}} R^2 \sim 10^{67}\ \text{cm}^3 \ , \qquad (21)$$

where we assumed that the radius of the disk $R \sim 15$ kpc $\sim 5 \times 10^{22}$ cm, $h_{\mathbf{d}} \sim 200$ pc, and the density of the cosmic rays

$$w_{\mathbf{c.r.}} \equiv w_{\mathbf{c.r.,G}} \sim 10^{-12}\ \text{erg/cm}^3 \ .$$

Disk models are, however, inapplicable. Indeed, in these models the cosmic rays must be contained in a disk with half-thickness $h_{\mathbf{d}} \ll R$, notwithstanding the high pressure of the cosmic rays

$$p_{\mathbf{c.r.}} = w_{\mathbf{c.r.}}/3 \sim 3 \times 10^{-13}\ \text{dyne/cm}^2$$

acting from within. Using as an example the devices for thermo-nuclear fusion and from theoretical considerations it is known how difficult it is to secure such a containment. To be concrete, the cosmic rays will leave the disk along the scattered magnetic field lines (even if we assume the magnetic field in the disk to be almost closed, this closure can under cosmic conditions not be ideal), drift because of the bending of the field lines and as the result of the development of instabilities. Most important, however, are the direct radio-astronomical data which indicate that cosmic rays, indeed, leave the gas disk. This is clear from the fact that rather bright galactic radio-emission is observed at distances from the galactic plane which are considerably larger than the half-thickness $h_d \sim 100$ to 200 pc. Thus, if we introduce the concept of a radio-disk, as is often done, where one understands by the radio-disk, the region with uniform emittance responsible for the whole of the galactic synchrotron radio-emission, the half-thickness of the radio-disk $h_r \sim 800$ pc. In practice, however, there are no grounds for assuming that cosmic rays and, in particular, the radiating relativistic electrons fill such a disk uniformly. It is natural to think, contrariwise, that the cosmic rays surround the gas disk and form some kind of corona or halo. In fact, S.V. Pikel'ner[19] put forward the assumption of the existence of such a 'cosmic ray halo' in 1953. If the magnetic field in the halo decreases with increasing distance from the galactic plane not too fast (say, approximately in such a way that the relation $w_{c.r.} \sim H^2/8\pi$ is retained) we must observe a 'radio-halo'; the electron component of the cosmic rays proceeding from the source region is responsible for the radio-emission of this halo. As the electrons, in contrast to the protons and nuclei, undergo synchrotron and Compton energy losses the size of the radio-halo (especially at relatively high frequencies) must be less than the size of the cosmic ray halo. The observation of a radio-halo would thus corroborate that there exists an even larger, powerful cosmic ray halo. On the other hand, the absence of a noticeable radio-halo can be accounted for even when there is a cosmic ray halo present by taking into account the possible fast decrease in the magnetic field strength in the halo and the selection of insufficiently low frequencies for the measurements. To establish that there is a radio-halo for our Galaxy is furthermore made very difficult in connection with the fact that we are inside the halo.

Somehow the problem of the radio-halo of our Galaxy has turned out to be hard to solve and its discussion has led to contradictory conclusions. However, at the present the problem is, I am convinced, very definitely cleared up both as the result of the analysis of the radio-data for our Galaxy and through the observation of other spiral galaxies, seen 'edge-on'

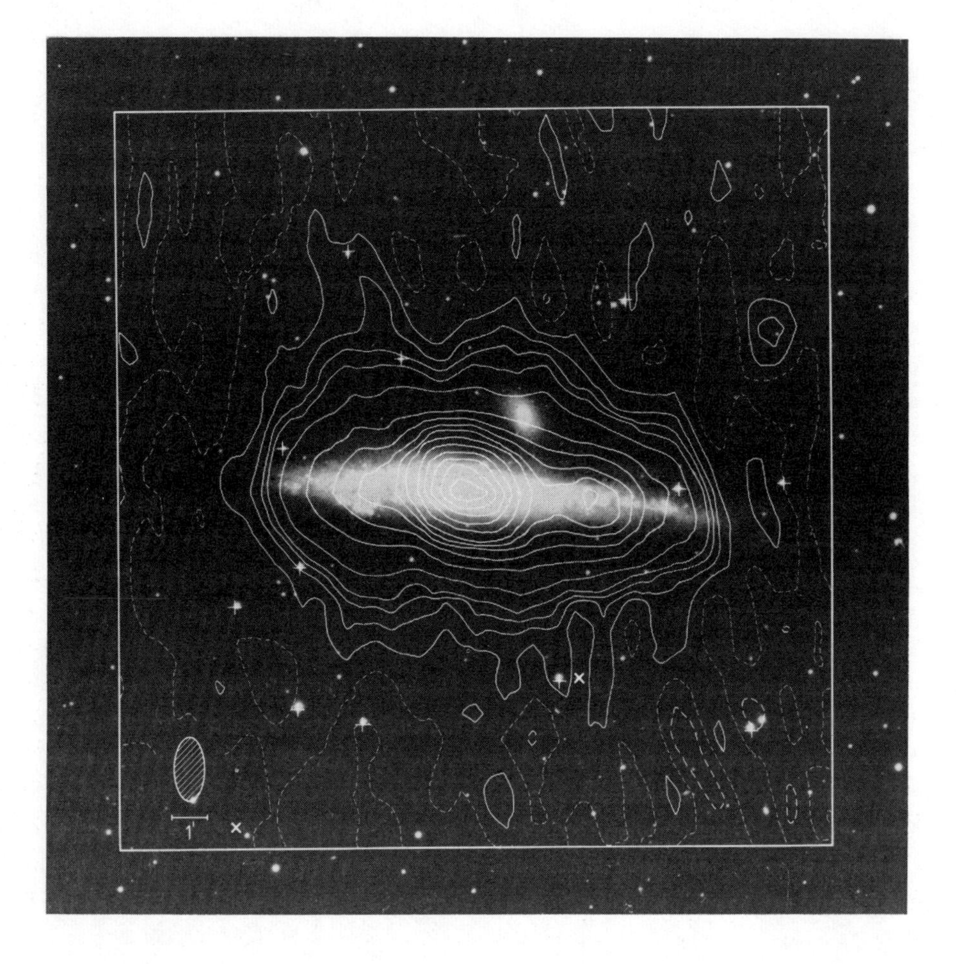

Fig. 3.1

(in the first instance I am thinking of the galaxies NGC 4631 and NGC 891). The corresponding data and references to the literature can be found in Refs. 12, 13, and 20. In Fig. 3.1 we show (with white lines) the radio-isophotes for the galaxy NGC 4631 at a wavelength $\lambda = 49.2$ cm (frequency $\nu = c/\lambda = 610$ MHz); the darkened regions are the optical image of the same galaxy, and from comparing them it is clear that there is a radio-halo present. At longer wavelengths the radio-halo must be even larger.

From what we have said it follows that only a galactic model with a halo is acceptable; this model had been applied already many years earlier (see, for instance, Ref. 18) but it was not directly substantiated because of the uncertainty in the problem of the existence of a radio-halo.

The characteristic dimensions of the cosmic ray halo are equal to the radius of the galactic disk $R \sim 5 \times 10^{22}$ cm. However, it is quite possible that the halo may turn out to be somewhat flattened with a characteristic half-thickness $h_h \sim 10$ kpc $\sim 3 \times 10^{22}$ cm. It is, of course, difficult to expect the presence of a sharp boundary of the halo and it is probable that when we go away from the galactic centre the magnetic field strength in the halo drops; the same will be true for the energy density $w_{c.r.}$. We thus understand by h_h just a characteristic dimension corresponding to a not too strong decrease in the density $w_{c.r.}$. The characteristic volume (confinement region) and the cosmic ray energy in the model with halo are thus as follows:

$$W_{c.r.} \sim w_{c.r.} V_h \sim 10^{56} \text{ erg} \quad , \quad V_h \sim (1 \text{ to } 5) \times 10^{68} \text{ cm}^3 \ , \tag{22}$$

where we have assumed that at the periphery of the halo the density $w_{c.r.}$ must decrease. From various considerations it follows that the lifetime of the proton-nuclear component of the cosmic rays in the system (in the halo), which is determined by the escape of the cosmic rays into intergalactic space, is $T_{c.r.} \sim (1 \text{ to } 3) \times 10^8$ years $\sim (3 \text{ to } 10) \times 10^{15}$ s.* Hence, in a quasi-stationary state (which in general corresponds to the actual situation) the power of cosmic ray generation in our Galaxy is:

$$U_{c.r.} \sim \frac{W_{c.r.}}{T_{c.r.}} \sim (1 \text{ to } 3) \times 10^{40} \text{ erg/s} \ . \tag{23}$$

---

* We give, for instance, the following estimate. In one-dimensional diffusion the particles traverse on average during a time $T_{c.r.}$ a path $L = \sqrt{(2DT_{c.r.})}$, where $D \sim lv$ is the diffusion coefficient (l is the mean free path and v the average velocity of the cosmic rays between 'collisions', that is, appreciable deflections in the magnetic fields). When $L \sim h_h \sim 3 \times 10^{22}$ cm and $D \sim 10^{29}$ cm^2/s ($l \sim 10^{19}$ cm ~ 3 pc and $v \sim 10^{10}$ cm/s) the time $T_{c.r.} \sim 3 \times 10^{15}$ s $\sim 10^8$ years. One can also arrive at the same estimates for $T_{c.r.}$ and D in other ways.

It is very important that the estimate (23) is practically independent of the parameters of the model (the time $T_{c.r.}$, the volume V, and the energy $W_{c.r.} = w_{c.r.}V$). The fact is that we know from the data about the chemical composition of the cosmic rays the average path x traversed by them in matter (in the interstellar medium) $x \approx 5$ g/cm^2. At the same time for ultra-relativistic particles (velocity $v = c$) $x = c\bar{\rho}T_{c.r.}$, where $\bar{\rho}$ is the average gas density in the confinement region and $T_{c.r.}$ the time spent in it. Clearly,

$$U_{c.r.} \sim \frac{W_{c.r.}}{T_{c.r.}} = \frac{cw_{c.r.}V\bar{\rho}}{x} = \frac{cw_{c.r.}M_g}{x} \sim 5 \times 10^{-3} M_g \sim 5 \times 10^{40} \text{ erg/s} ,$$

where $M_g = \bar{\rho}V$ is the mass of the gas in the Galaxy which we take to be approximately equal to 10^{43} g, using independent astronomical data (when we take the contribution from molecular hydrogen H_2 into account this estimate may need an increase by a factor 2 or 3). Of course, the accuracy of all such estimates given here is not so high that we can attach any importance to a discrepancy by a factor 3 to 5.

For the electron component of the cosmic rays in our Galaxy

$$W_{c.r.,e} \sim w_{c.r.,e} V_h \sim 10^{54} \text{ erg} , \quad U_{c.r.,e} \sim \frac{W_{c.r.,e}}{T_{c.r.,e}} \sim 10^{39} \text{ erg/s} . \quad (24)$$

We have assumed here that $w_{c.r.,e} \sim 10^{-2} w_{c.r.}$, and that the effective time $T_{c.r.,e} \sim 10^{15}$ s is somewhat shorter than $T_{c.r.}$ because of the presence of losses. The losses are mainly synchrotron losses due to which the value of $U_{c.r.,e}$ must be close to the total power of the galactic radio-emission; this is, indeed, the case. The power (23) of cosmic ray generation must be admitted to be rather large, if we bear in mind that the total luminosity of the Sun $L_{\odot} = 3.8 \times 10^{33}$ erg/s, while the luminosity of the whole Galaxy is $L_G \sim 10^{44}$ erg/s. We have already mentioned that supernovae may, if we consider energy possibilities, secure the acceleration of all cosmic rays. For example, the energy release in a single supernova outburst is, on average, assumed to be equal to 10^{49} to 10^{51} erg. In our Galaxy supernovae occur on average every 10 to 30 years. Hence it follows that the average power for supernovae is $U_{sn} \sim 10^{40}$ to 3×10^{42} erg/s. Acceleration of the cosmic rays with a power $U_{c.r.} \sim 1$ to 3×10^{40} erg/s is thus possible. Analogous estimates, which are even more reliable because of the use also of radio-astronomical data, indicate the possibility of accelerating electrons to the power (24).

It is altogether clear that a favourable energy estimate is necessary for a choice of sources, but not yet sufficient. It is appropriate to mention once again that far from all stars can guarantee the power for generating cosmic rays (23). For instance, the average power of cosmic ray generation by the Sun $U_{c.r.} \sim 10^{25}$ erg/s. Therefore, even 10^{11} stars of the same type as the Sun (the mass of the Galaxy $M_G \sim 10^{11} M_\odot$) would secure only a generation with a power 10^{36} ergs which is four to five orders of magnitude less than the necessary power (23). Hence, it is in any case not easy to secure the necessary generation power, using non-exploding stars (O-type stars, and so on).

We assume therefore that the parameters for the model for the origin of cosmic rays which corresponds to reality – the galactic model with a halo – is to a first approximation established (see (23) and (24); the half-width of the halo $h_h \sim 10$ kpc $\sim 3 \times 10^{22}$ cm).

The value of a reliable choice of model can hardly be overestimated, especially in the light of the history of the problem. At the same time, choosing a model means only to lay the foundations for that edifice which we wish to construct. To do this, and in order that we can assume the problem of the origin of the cosmic rays observed at the Earth to be more or less solved, it is on the whole necessary to give answers to a number of questions and to solve many problems. Let us list the basic problems and questions.

(1) There is still little known about the configuration and strength of the magnetic field in the halo, and indeed, about the parameters in general of the cosmic ray halo. Even apart from the fact that a knowledge of the structure of the field in the halo is very important for an explanation of the origin of particles with very high energies, it is necessary for a quantitative analysis of the radio-data and for a consideration of the departure (diffusion) of cosmic rays from the Galaxy. Progress is possible in this respect as a result of measurements of the polarization of the radio-emission and, possibly, of measurements of the anisotropy of cosmic rays with very high energies.

(2) At present one may assume that the diffusion model with a halo is the best or, at any rate, the most reasonable approximation. In these models we assume that the diffusion coefficient is a free parameter (or a number of free parameters, if we use different diffusion coefficients in the disk and the halo, anisotropic diffusion, and so on). It is necessary to specify the corresponding model in more detail on the basis of an analysis of the data about the chemical and isotopic composition of the cosmic rays and about their anisotropy (see Refs. 9 and 11 to 13 and the literature cited there).

(3) Even, if the diffusion model turns out to be completely satisfactory for a quantitative description of the chemical and isotopic composition (especially, for radioactive nuclei such as ^{10}Be at various energies) and also for the calculation of the anisotropy, the next step is unavoidable – the evaluation of the diffusion coefficients $D(E)$. To be precise, we are thinking

about establishing a connection between D and the parameters of the interstellar medium (including here data about the magnetic field). We must then bear in mind that diffusion in a magnetic field is naturally different from the diffusion of molecules in a gas. What is more, the wandering of cosmic rays, caused not only by scattering by magnetic field inhomogeneities but also by the divergence and entanglement of field lines, and also by convective motions in the interstellar medium, may under some circumstances be reduced to diffusion — in the sense of an applicability of the diffusion equation. However, in a number of cases (especially, if there is a galactic wind or convection on a very large length scale) the transfer equation must be supplemented by a convective term. It is natural that the explanation of the rôle of convection using cosmic ray data, but using also all other available information is part of the problem of cosmic ray astrophysics. Earlier we assumed, albeit tacitly, that the cosmic rays filled the disk and the halo more or less uniformly. At the same time there are certainly regions with higher cosmic ray densities — the supernova shells. It is quite possible and even probable that otherwise — except for relatively small regions around the sources — the cosmic ray density is really more or less uniform. The next step was the assumption about the approximate constancy of the densities $w_{\mathrm{c.r.}}$ and $w_{\mathrm{c.r.,e}} = \kappa_{\mathrm{e}}^{-1} w_{\mathrm{c.r.}}$ for the whole Galaxy (except small regions near the sources and, of course, restricting ourselves to not too large distances from the galactic centre, as the density $w_{\mathrm{c.r.}}$ and especially $w_{\mathrm{c.r.,e}}$ must decrease towards the 'boundaries' of the halo). There is, however, no doubt that one needs a confirmation of such an assumption which, in principle, can be made by radio- and gamma-astronomical methods.

(4) Because there are losses, the propagation (diffusion) of the electron component is an independent problem which one must solve, of course, taking into account all other data, especially those obtained for the proton-nuclear component. Whereas for the latter the comparison with observations reduces to an analysis of the chemical and isotopic composition and of the anisotropy, as well as of the energy spectra $I_{Z,A}(E)$, for the electron component there is, apart from the energy spectrum $I_{\mathrm{e}}(E)$ and, in principle, also the anisotropy $\delta_{\mathrm{e}}(E)$, already a rich radio-astronomical source of information which will grow in the future. Moreover, gamma-astronomical measurements can give information about the softer part of the electron spectrum; there are also some indirect methods for studying the electron component. The problem consists in constructing a self-consistent model which will allow us to connect and explain all above-mentioned data about the electrons. We may especially

mention the necessity for a detailed study (survey) of the radio-halo at different frequencies.

(5) The positrons in the cosmic rays are, very probably, practically all secondary particles — they are formed in the interstellar gas or in the source itself in collisions of the cosmic rays. It is true that the existence of a 'primary' positron component, for instance from pulsars, is not excluded, but this is a special problem. Assuming that the positrons are secondary particles one can evaluate their intensity at the Earth $I_{e^+}(E)$, if we give a definite model for the origin of the cosmic rays and use the data for the interstellar medium. As a result a comparison of such calculations with measurements of $I_{e^+}(E)$ (so far they are as yet very incomplete; see section 2) must serve the cause of verifying the assumed model. For this purpose one must also use the measurements of the antiproton component $I_{\bar{p}}(E)$ (which have already started recently).

(6) We have already briefly discussed the problem of the sources. There is no question of an anything like complete picture unless we clarify the problem of the share of the various sources. Close to this, although it is in a certain sense independent, is the great problem of the acceleration and injection mechanisms for cosmic rays.

(7) A special place is occupied by the problem of cosmic rays with very high energies ($E \geqslant 10^{17}$ eV) observed at the Earth, and their origin.[10] In this range, and especially for $E \geqslant 10^{19}$ eV, a metagalactic origin is not excluded at all and is even rather probable. If this is, indeed, the case, particles with very high energies are the only extra-galactic charged particles reaching the Earth.

(8) So far we have concentrated on cosmic rays in our Galaxy. However, for cosmic ray astrophysics as a whole our Galaxy takes up a special place only because of the position of the solar system. It is inappropriate to dwell here in detail on problems of high-energy extra-galactic astrophysics. It is sufficient to mention the particular large rôle played by cosmic rays in radio-galaxies and quasars. Information about extra-galactic cosmic rays can be obtained by radio-, x-ray-, and gamma-astronomical methods. High-energy neutrinos are produced by cosmic rays and these neutrinos have both a galactic and an extra-galactic origin. It is impossible to refrain from mentioning separately the cosmic rays in intergalactic space and, in particular, in clusters of galaxies and in the Local Supercluster. The study of intergalactic cosmic rays is especially difficult because of the fact that the intergalactic field is weak, that the gas density is low, and that the energy density

of the cosmic rays and their electron component is small.* However, it is probable that for dense clusters one can observe a radio-halo of the clusters, similar to the radio-halo of individual galaxies. The background (diffuse) Metagalactic electromagnetic radiation in various frequency ranges is also measurable (it is another question that such radiation in some cases reduces to the joint emission of discrete sources — this problem is not yet clear).

The selection of some eight problems from cosmic ray astrophysics, which has been made here, is, of course, rather arbitrary — this selection could have been made differently and in more detail. Moreover, we have not mentioned at all the solar cosmic rays, the modulation of the cosmic rays, the radiation belts of the planets (Earth, Jupiter, and a few more). Nonetheless, I feel that it is clear already from what I have said how broad the problematics involved in cosmic ray astrophysics is.

V. THE PERSPECTIVES FOR THE DEVELOPMENT OF COSMIC RAY ASTROPHYSICS

At this moment cosmic ray astrophysics enters and, in fact, has already entered a new stage in its development. Firstly, the galactic model with a halo has finally been proved — this is, at any rate, my opinion. This does not mean that, for instance, there is no need for a more detailed study of the statement that the energy density $w_{c.r.}$ decreases when we go away from the Sun in the direction of the anticentre. However, in my opinion we are dealing only with the filling in of details and some reasonable reservations, but basically all has been done. Secondly, in a number of cases new apparatus and devices are being prepared or planned, and some of this has already been started, which will enable us to obtain in the foreseeable future a rich harvest of information about the primary cosmic rays and the electromagnetic and neutrino emission produced by them.

This is the general situation in cosmic ray astrophysics. Bearing this in mind it is appropriate to finish this lecture with a discussion of the perspectives for the development of cosmic ray astrophysics.

It is impossible, of course, to give a detailed and complete prognosis of the development of cosmic ray astrophysics. However, the work of the past (cosmic rays have now been studied for already 70 years and the astrophysics of cosmic rays for about 30 years) enables us to foresee quite a lot. Moreover, the complexity of modern apparatus has led to the fact that in a number

---

* We start here, of course, from the assumption that the galactic model of the origin of the cosmic rays which are observed at the Earth is correct (*vide supra*).

of cases the planning and construction of new devices will take more than one year. That is the reason why, to give an idea about the character of the development of cosmic ray astrophysics in the eighties, and to some extent also up to the end of the century and even up to the centennial jubilee of the discovery of cosmic rays*, is not utopia.

It seems to me that by 2001 or at any rate by 2012 we may expect to see a solution of almost all the problems formulated at the end of the preceding section. There will, of course, arise new problems and there will be unexpected results. It is hardly possible to give a more precise prognosis when we are dealing with such a relatively long period of 20 to 30 years. As far as the immediate future is concerned, that is, the present decade, and, possibly, the start of the next one, one can note more concrete expectations and possibilities.

(1) One may expect a determination of the chemical composition and the energy spectrum of the nuclei up to energies of 10^{12} to 10^{13} eV/nucleon. The isotopic composition becomes known only for lower energies. At the same time the determination of the isotopic composition for a number of stable nuclei even at low energies $\varepsilon_k \sim 10^9$ eV/nucleon is very important for solving the problem of the sources and for making more precise the peculiarities of the propagation and fragmentation of the cosmic rays in the interstellar medium. We must specially mention the study of radioactive nuclei. To be concrete, one may hope soon to obtain for the nucleus ^{10}Be with an average lifetime $\tau = 2.2 \times 10^6\ E/Mc^2$ years information about the number of such nuclei in cosmic rays with $E/Mc^2 \leqslant 10$. A reliable measurement of this number would be a great achievement.

The working out of the data which we have here in mind would enable us probably to advance further along the road to filling in the details of the galactic model for the origin of the cosmic rays.

(2) The cosmic ray spectrum has at the present been traced up to energies around 10^{20} eV. The number of particles for very high energies $E \gtrsim 10^{16}$ to 10^{17} eV is so small† that we can observe them only in extensive showers in the atmosphere. The chemical composition of cosmic rays with very high

---

* As the date of the jubilee one should really give 7th August 2012, as the most successful flight undertaken by Hess took place on 7th August 1912.

† The intensity of primary particles with energies $E \geqslant 10^{16}$ eV is not more than 10^2 particles/km^2 sterad hour. When $E \gtrsim 10^{20}$ eV this intensity has already dropped to the order of 10^{-6} particles/km^2 sterad hour $\sim 10^{-2}$ particles/km^2 per sterad per year.

energies is completely insufficiently clarified, and their origin is alsoo obscure. Neither a galactic nor an extragalactic origin (we think here of the Local Supercluster; particles with the observed energy spectrum cannot come from regions which are further away because of losses through the interaction with the relict radiation) is now excluded. One may hope that the problem will be solved (in the sense of a provisional localization of the sources and so on) within 10 to 20 years (one of the most important trends of research is the measurement of the anisotropy of the cosmic rays with very high energies).

Apart from having astrophysical importance, the study of cosmic rays with very high energies retains and, in all probability, for a long time will continue to have its importance for physics. We remind ourselves that in the period 1927 to 1929 and up to the start of the fifties cosmic rays were widely used in high-energy physics and through its means the following particles were discovered:[1] the positron, e^+ (1932), the $\mu^{\pm}$-mesons (1937), the $\pi^{\pm}$-mesons (1947), the K^0- and $K^{\pm}$-mesons (1947-48) and the Λ, Σ^+, and Ξ^--hyperons (1951-53). However, since then the centre of gravity of the corresponding physical investigations has shifted to accelerators. This is very understandable and, if one can for a given energy E use an accelerator, cosmic rays are unable to compete. However, in the eighties one can, apparently, when using storage rings with colliding beams of protons or of protons with antiprotons, to aim at most at energies of $E_c = 10^{12}$ eV = 1 TeV in each beam. When we change from the centre of mass frame to the laboratory frame this is equivalent to using protons with energies

$$E = 2(E_c)^2/Mc^2 \approx 2 \times 10^{15} \text{ eV}.$$

Thus, for energies $E > 2 \times 10^{15}$ eV, or, if we consider an accelerator with $E_c = 3$ TeV for $E > 10^{16}$ eV the only particle source, up to the end of the century will be the cosmic rays. Of course, it is very difficult to work with cosmic rays in the energy range $E > 10^{15}$ to 10^{16} eV with the aim of performing physical investigations, but it was always difficult to do research and a new technique, which our predecessors did not know about, may enable us to cope with new difficulties.

(3) There are grounds for hoping that we can obtain the spectral $I_{e^+}(E)$ and $I_{\bar{p}}(E)$ for the positron and antiproton components. The spectrum $I_e(E)$ for the electron component (or, more precisely, the whole of the electron-positron component) at the Earth is already now known to a certain approximation. However, it is very important to know in more detail the way I_e

depends on the energy $E = E_e$ and this in possibly a wider range of energies. One may count on a successful solution of this problem through the use of new apparatus.

(4) The improvement of radio-telescopes proceeds. A few years ago this led to the observation of radio-halos for a number of spiral galaxies. However, this made the necessity of using longer wavelengths ($\lambda \gtrsim 1$ m) and to carry out polarization measurements to be felt even more acutely. When the corresponding results will have been obtained, and we can reckon on this in the foreseeable future, it will be possible to progress appreciably in our study of the radio-halo and the cosmic ray halo of our Galaxy and of other galaxies and, possibly, of galactic clusters.

(5) Earlier we made a number of remarks regarding gamma-astronomy. Apparently, one may assume that also in this respect we experience a turning point. The present decade will, probably, lead to progress which is similar to that which occurred in the seventies in X-ray astronomy (in this field one may consider the results obtained by the 'Einstein' observatory to be the culmination). The new generation of gamma-telescopes will give us an opportunity not only to refine the results from the SAS-2, COS-B, and other satellites, but also to study a large number of discrete sources, among them the Magellanic Clouds, a number of galaxies and their nuclei, quasars. What has been obtained already [12] is rather impressive, for instance, the result that the gamma-luminosity of the quasar 3C 273 is equal to

$$L_\gamma(50 < E_\gamma < 500 \text{ MeV}) = 2 \times 10^{46} \text{ erg/s}$$

(assuming a distance $R = 790$ Mpc). Over a period of 10^6 years such a luminosity corresponds to an emission of an energy

$$W_\gamma \sim 6 \times 10^{59} \text{ erg} \sim 3 \times 10^5 M_\odot c^2$$

in gamma-photons alone. The optical and X-ray luminosity of this quasar is approximately the same as the gamma-luminosity and only in the infrared region is the luminosity higher by an order of magnitude. For the pulsar PSR 0532 (in the Crab)

$$L_\gamma(E_\gamma > 100 \text{ MeV}) \approx 3.5 \times 10^{34} \text{ erg/s}.$$

For the source Cyg X-3 (which may be a young pulsar in a binary system)

$$L_\gamma(E_\gamma > 40 \text{ MeV}) \approx 7.5 \times 10^{36} \text{ erg/s},$$

$$L_\gamma(E_\gamma > 10^{12} \text{ eV}) \approx 1.0 \times 10^{35} \text{ erg/s}.$$

Large gamma-luminosities are very impressive and, at any rate, indicate the presence of a large number of cosmic rays (electrons, if we think about the Compton mechanism, and also about synchrotron and curvature radiation in the case of pulsars). We may also expect many interesting results in the rather near future from other branches of gamma-astronomy (line emission, and so on). High-energy astrophysics is closely connected with the studies in the X-ray and, in some cases, in other bands. It is, however, very clear that the development occurs along a broad front and here we wish merely to pick out a few of the main points.

(6) High-energy neutrino astrophysics belongs to these main points. This field has only just started, if we are thinking about experiments. However, already fully realistic are underground neutrino measurements, say, from supernova outbursts in our Galaxy. The construction of deep-water optical and/or acoustical systems (DUMAND project and others) enables us to fix confidently with a rather high level of angular resolution of the order of 1° neutrinos with energies $E_\nu \gtrsim 10^{12}$ eV from galactic or far extra-galactic sources. We have already mentioned that neutrinos of this energy are produced practically only by the proton-nuclear component of the cosmic rays and that they therefore can serve as its indicator (this is similar to the gamma-rays from π^0-decay in the softer region of the spectrum). Moreover, the neutrinos possess an extremely high penetration power, while even gamma-rays with energies $E_\gamma \gtrsim 2 \times 10^{11}$ to 10^{14} eV at large metagalactic distances are already strongly absorbed by electromagnetic radiation (the $\gamma + \gamma' \rightarrow e^+ + e^-$ process, where the soft photons of the relict and the optical radiation play the rôle of γ'). Gamma-rays are all the more absorbed in a layer of matter which exceeds 100 g/cm^2. They cannot escape, therefore, from the interior regions of, say, dense galactic nuclei. To a large extent the difficult of explaining the nature of the cores of quasars and of the active nuclei of galaxies is just connected with this fact. The detection of neutrino emission from these objects, in conjunction with gamma-astronomical observations opens up for some models the possibility to distinguish between a massive black hole and a magnetoid.[21]

In general, one may think that high-energy neutrino astronomy ($E_\nu > 10^{11}$ eV) * is as yet the most important of the unused reserves of high-energy astrophysics and of astronomy as a whole (it is true that in the last

---

* Low-energy neutrino astronomy ($E_\nu \lesssim 10$ to 20 MeV) is, of course, also one of the most important directions of research for a study of the Sun, of supernova outbursts and, possibly, of some other objects (see also the 'key problems' lectures).

case it is impossible to refrain from mentioning also the not less important 'reserve' gravitational wave astronomy).

(7) Concluding this brief catalogue of trends of research and possibilities for further development we must also mention theory. It is difficult to overestimate the rôle of theory in physics, but in astrophysics it is somehow even larger because of the difficulty of obtaining observational data and the impossibility to do experiments beyond the solar system. In cosmic ray astrophysics the theory has always had and still has a prominent position; there are no reasons whatever for expecting this position to change in the future. The nature of the topical problems which are in front of the theory is, I may hope, clear from what we have said so far.

Cosmic ray astrophysics has already reached a certain maturity and today this field is very broad and has many branches. At the present time a new stage has started, as we have already emphasized, for the development of cosmic ray astrophysics; it is characterized by a transition to making the galactic model with a halo more precise in a quantitative sense and by the appearance of new methods of investigation and many new possibilities.

REFERENCES

1. A.M. Hillas, Cosmic Rays, Pergamon Press, Oxford, 1972.
2. *Selected Papers on Cosmic Ray Origin Theories* (Ed. S. Rosen), Dover Publications, New York, 1969.
3. *Early History of Cosmic Ray Studies* (Eds. Y. Secido and H. Elliott), D. Reidel, Dordrecht, Netherlands, 1982.
4. W. Baade and F. Zwicky, Proc. Nat. Acad. Sc. USA , **20** (1934) 259; Phys. Rev., **46** (1934) 76.
5. E. Fermi, Phys. Rev., **75** (1949) 1169
6. *Paris Symposium on Radio Astronomy*, Standford University Press, 1959.
7. V.L. Ginzburg and S.I. Syrovatskii, Ann. Rev. Astron. Astrophys., **3** (1965) 297.
8. V.L. Ginzburg, *Theoretical Physics and Astrophysics*, Pergamon Press, Oxford, 1979.
9. V.L. Ginzburg and S.I. Syrovatskii, *The Origin of Cosmic Rays*, Pergamon Press, Oxford, 1964.
10. G.D. Rochester and K.E. Turver, Contemp. Phys., **22** (1981) 425.
11. V.L. Ginzburg and V.S. Ptuskin, Rev. Mod. Phys., **48** (1976) 161,675.
12. *Fifteenth International Cosmic Ray Conference (ICRC)*, Conference Papers, Plovdiv, Bulgaria, 1977;
    Sixteenth ICRC, Conference Papers, Kyoto, Japan, 1979;
    Seventeenth ICRC, Conference Papers, Paris, France, 1981.
13. *Origin of Cosmic Rays*, IUPAP/IAU Symposium No.94, Bologna, Italy, D. Reidel, Dordrecht, Netherlands, 1981.
14. V.L. Ginzburg, Phil. Trans. Roy. Soc., **A277** (1975) 463.
15. G. Burbridge, Phil. Trans. Roy. Soc., **A277** (1975) 481.
16. V.L. Ginzburg, Nature (Phys, Sc.), **239** (1972) 8.
17. D. Dodds, A.W. Strong, and A.V. Wolfendale, Monthly Not. Roy. Astron. Soc. **171** (1975) 569;
    A.V. Wolfendale, Pramana (India), **12** (1979) 631.

18. V.L. Ginzburg, Usp. Fiz. Nauk., **51** (1953) 343; Fortschr. Physik, **1** (1954) 659.
19. S.B. Pikel'ner, Doklady Akad. Nauk. SSSR, **88** (1953) 229.
20. V.L. Ginzburg, Sov. Phys. Uspekhi, **21** (1978) 155.
21. V.S. Berezinsky and V.L. Ginzburg, Monthly Not. Roy. Astron. Soc., **194**, (1981) 3.

CHAPTER 4

Radiation by a Uniformly Moving Charge

(Cherenkov Effect, Transition Radiation, and Related Phenomena)

1. INTRODUCTION

There exists a somewhat isolated section of electrodynamics – and, in particular, of optics – which is dealing with the radiation by uniformly moving charges and other sources. Of course, if the source has an eigenfrequency which is different from zero we are dealing with the Doppler effect and when the motion is in vacuo this case is considered in all textbooks. However, when the source moves in a medium even the Doppler effect may turn out to be appreciably more complicated than in vacuo – and this is, as a rule, not yet reflected in the treatment in textbooks. Yet more important is the fact that uniformly moving sources (charges, dipoles, and so on) can radiate, even when they do not possess any eigenfrequency – that is, if in the frame of reference in which they are at rest, they are static. This is the case for the Cherenkov effect* and for transition radiation in its different variants.

This class of problems turned out to be, as we said, to some extent isolated and relatively badly known for the reason that it only rather recently has been discussed widely, even though it could have been considered many decades earlier. In turn this delay to a large extent is explained by the dogma which was prevalent in electrodynamics and which can be formulated as follows: a uniformly moving charge does not radiate.

---

* In the USSR this effect is called the Vavilov-Cherenkov effect to indicate the important rôle played by Vavilov in its discovery. In the English literature, however, it is usually called the Cherenkov effect, and we shall for the sake of brevity use this nomenclature in what follows.

It would seem that such a statement is clearly true. Indeed, if a charge, according to our assumption, moves uniformly (in a given inertial frame of reference) one can change to a frame of reference in which the charge is at rest. However, a charge at rest cannot radiate — it is sufficient to say that it does not have the energy necessary for the radiation. Finally the absence of radiation in the rest frame means that there is also no radiation in other inertial frames of reference.

However, what we have said is even for the case of the motion of a charge in vacuo valid only under two important provisos. Firstly, the motion must be uniform and rectilinear (the velocity $\vec{v}$ = constant) at all times, that is, in the time interval $-\infty < t < \infty$. If, however, the charge had been accelerated in the past — and this is practically inevitable — its field must in some cases 'relax' over a rather long time — that is, go over into a stationary state and this, in general, is accompanied by radiation.* Secondly, there is no radiation only if the velocity of the source is less than the speed of light $v < c = 3 \times 10^{10}$ cm/s. At the same time, even if we forget about the hypothetical tachyons which, probably, cannot exist (for them $v > c$) radiation sources (it is true that this is not the case for separate particles such as electrons) can well move with superluminal velocities $v > c$ (*vide infra*)

If, however, a charge moves uniformly not in vacuo, but in a medium, one must assume that the absence of radiation is the exception rather than the rule. In fact, a charge does not radiate only when its velocity v is less than the phase velocity c_{ph} for any of the electromagnetic waves which can propagate in the given medium. When $v > c_{ph}$ Cherenkov radiation occurs. Moreover, even if $v < c_{ph}$ absence of radiation refers only to a medium at rest, which is everywhere uniform and, moreover, which does not change with time. In a non-uniform and/or non-stationary medium a uniformly moving charge radiates also when $v < c_{ph}$ — this is transition radiation. It is interesting that even a charge at rest can radiate in a medium, although it is true that this is a rather exotic case (see section 5 below).

In the following we shall briefly discuss the radiation of sources (in the first instance, a charge) which moves uniformly in a medium. We shall fix our attention on physical ideas and results; the reader can find detailed

---

* For some details we refer to chapter 1 of Ref. 1 and the literature cited there. As this book is easily accessible and contains a rather broad bibliography I shall not give here references to the original literature. For the case of transition radiation and transition scattering we shall refer, apart from to Ref. 1, only to a recent review article [2] which also contains a bibliography.

calculations and references in the literature in Refs. 1 and 2.

II. THE CHERENKOV EFFECT

The possibility that a uniformly moving charge may radiate had in some form or other been considered long before the discovery and explanation of the Cherenkov effect (1934 to 1937). To be concrete, this problem had been discussed by Heaviside (1888 and later), Kelvin (1901), and Sommerfeld (1904). For instance, Sommerfeld considered in some detail the following problem: a charge (charged pellet) moves in vacuo with a velocity $\vec{v}$ = constant; find its electromagnetic field. When $v < c$ there is no radiation (a stationary problem is considered). When $v > c$ there occurs radiation and its front forms a conic surface with an angle θ_0 between the normal to it (the wave-vector $\vec{k}$) and the velocity $\vec{v}$. In that case (Fig. 4.1)

$$\cos\theta_0 = \frac{c}{v}. \tag{1}$$

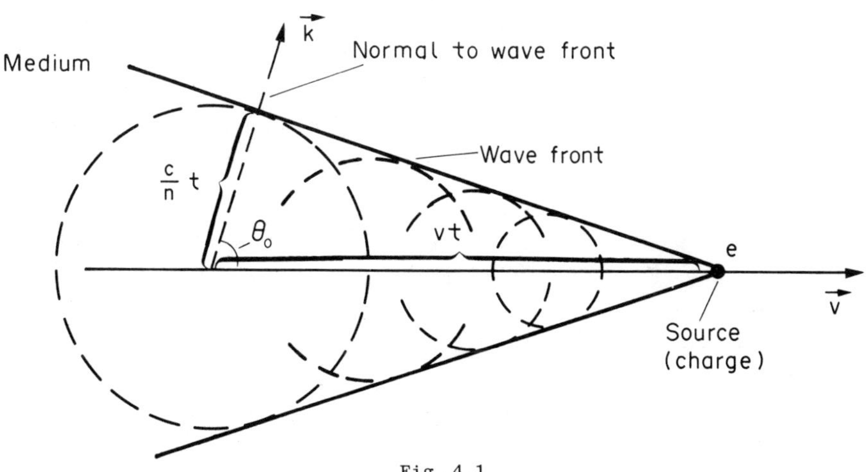

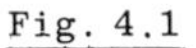
Fig. 4.1

One may say that condition (1) has a kinematic character — it is the condition for the interference of secondary waves which are excited by the charge along its path. To be more precise, under condition (1) waves which in the given case have a phase velocity $c_{\mathrm{ph}} = c$ (vacuum) and which are emitted along the trajectory of the source are in phase on the appropriate conical surface (this is clear from Fig. 4.1; after some time t the charge has traversed a path vt, and the wave a path ct if $n = 1$). Such a picture corresponds to the Huygens principle and is, of course, especially realistic when we are dealing with any kind of wave in a medium, but it is well known that the result is also true for a vacuum, notwithstanding the fact that there is no ether.

Rewriting it in the form

$$\cos \theta_0 = \frac{c_{ph}}{v} , \tag{2}$$

it is clear from what we have said that condition (1) refers to any kind of wave and for any static sources, but it can be satisfied only if

$$v \geqslant c_{ph} . \tag{3}$$

In acoustics condition (2) with $c_{ph} = u$, the velocity of sound, had been known for a long time (we are dealing here with the Mach condition and, correspondingly, with the Mach cone for supersonic sources — bullets, missiles, and so on).

As Sommerfeld solved the electrodynamic problem he obtained, of course, condition (1), so to say automatically, and he evaluated the intensity of the radiation. Of course, it is non-vanishing only when $v > c$ (see (1) to (3)). The formula obtained by Sommerfeld corresponds to the Tamm-Frank formula for the intensity of the Cherenkov radiation (see formula (6) below) in the case of a non-dispersive medium — the vacuum (index of refraction $n(\omega) = 1$).

In some sense Sommerfeld was 'unlucky' — within a year (1905) the special theory of relativity appeared. According to the latter the momentum of a particle is $\vec{p} = m\vec{v}/\sqrt{[1-(v^2/c^2)]}$ and it is impossible to accelerate it from small velocities to a velocity $v > c$. Moreover, it would appear that causality requirements also do not allow to have particles moving with a velocity $v > c$ as they could be used as superluminal 'signals'. It is true that relatively recently people have started to consider superluminal particles (tachyons, for which $v > c$) but I feel altogether that there existence is impossible (in first instance in connection with the requirement that causality must be observed), even apart from the fact that there are no indications whatever of an experimental nature in favour of the tachyon hypothesis. At any rate at the beginning of the century nobody at all was thinking of such tachyons as far as I know and Sommerfeld's paper was for many years forgotten.

It appears that for a long time nobody thought of the simple idea of transferring Sommerfeld's results to the case of the motion of a charge through a medium.* History developed differently. In 1934 S.I. Vavilov and P.A. Cherenkov observed the Cherenkov radiation, the nature of which was explained in 1937 by I.E. Tamm and I.M. Frank.

---

* Apparently Heaviside[3] correctly understood both the rôle of the medium and the rôle, discussed below, of superluminal 'light spots'. However, in his time his papers did not attract the necessary attention. I have not investigated Heaviside's papers in more detail, although this would be rather interesting (see T.R. Kaiser, Nature, **247** (1974) 400).

The Cherenkov radiation (I am using here the Western terminology, although I and many people in the USSR feel that it would be more correct to speak of Vavilov-Cherenkov radiation) is the radiation of a uniformly moving charge in a transparent medium with refractive index $n(\omega)$ and, hence, with phase velocity $c_{ph} = c/n(\omega)$. The conditions (2),(3) for emission thus become

$$\cos\theta_0 = \frac{c}{n(\omega)v} \tag{4}$$

and

$$v \geqslant \frac{c}{n(\omega)} . \tag{5}$$

Tamm and Frank* obtained for the energy emitted by a particle of charge e per unit time (that is, along a path length v) the following expression:

$$\frac{dW}{dt} = \frac{e^2 v}{c^2} \int_{\frac{c}{n(\omega)v} \leqslant 1} \left(1 - \frac{c^2}{n^2(\omega)v^2}\right) \omega\, d\omega = \frac{e^2 v}{c^2} \int_{\frac{c}{n(\omega)v} \leqslant 1} \sin^2\theta_0\, \omega\, d\omega . \tag{6}$$

Cherenkov radiation now occupies a prominent place in physics, and a huge number of papers is devoted to it, including books and review articles (see Refs. 1, 2 and the literature cited there). Not a small, but, perhaps, the main part is played here not by the Cherenkov effect in the proper sense of the word (optical emission by a charge which moves uniformly in a medium with velocity $v > c/r(\omega)$), but by ideas and analogues which are connected with it, one might well say, by a 'Cherenkov ideology'. As a typical example we give the interpretation of the so-called Landau damping, or, to be precise, the damping of longitudinal (plasma) waves in a collisionless plasma. L.D. Landau concluded in 1946 that such a damping should exist when he solved the problem with initial conditions of the propagation of longitudinal perturbations in a collisionless plasma using the kinetic equation. The collisionless damping which then occurs and which also appears in a number of other problems in plasma physics and plasma-like media (for instance, in the case of the 'solid-state plasma' – the electron liquid in metals, and so on) can be interpreted (if one considers it from a physics point of view) in different ways. One of them is the following: the condition for collisionless absorption of a wave by the electrons in the plasma, which has the form

$$\omega = (\vec{k}\cdot\vec{v}) , \tag{7}$$

is simply the condition for emission (2) for waves (in this case the longitudinal plasma waves) with a phase velocity

$$c_{ph} = \frac{\omega}{k} = \frac{c}{n_\ell} , \quad n_\ell(\omega) = \frac{ck}{\omega} , \tag{8}$$

---

* Tamm and Frank knew about the work of Sommerfeld only after they had completed their investigation.

where $n_\ell(\omega)$ is the refractive index for the longitudinal waves considered (as the velocity of the longitudinal waves c_{ph} is independent of c the introduction of the refractive index $n_\ell(\omega)$ has, it is true, rather a formal character). The collisionless Landau absorption is thus closely connected with the inverse Cherenkov effect for plasma waves (the kinematic conditions for absorption and for emission of waves when recoil is neglected are the same). Of course, when we are dealing with an 'external' wave (in this case a longitudinal wave propagating in the plasma) we must consider its interaction not only with a single particle, but with a collection of them. As a result it is necessary to take into account not only the absorption of waves, but also their stimulated (induced) emission.

It is just this explanation which is now, apparently, most widespread and makes collisionless absorption for many people completely obvious and understandable (one should, of course, not take what we have said to mean a denial of the complete validity of other 'intuitive' pictures, that is, of another physical language).

In general it is rather natural that the development of the physics of superluminal sources (sources which are moving with a velocity larger than the speed of light in the given medium) has proceeded and, I am certain, will proceed even further. We give here one example, which is a relatively recent one, but which has its roots in the rather far-distant past. Indeed, we have already emphasized that Sommerfeld's work was assumed not to bear any relation to reality as it is impossible to satisfy the requirements (3),(5) in vacuo. In accordance with this, up to recently it seemed rather obvious that Cherenkov radiation would turn out to be impossible in vacuo and also in media with a refractive index $n(\omega) < 1$ (in particular, in an isotropic plasma in the region where the well known formula $n^2(\omega) = 1 - \omega_p^2/\omega^2$ is valid). In actual fact, however, such a conclusion is incorrect, as I pointed out in 1972, or, if you like, is too rash. Apart from the above-mentioned but, in all probability, non-existing 'superluminal' particles, the tachyons, completely realistic radiation sources can move with a velocity $v > c$. Moreover, such sources have been known for a long time – as an example I may mention patches of light ('light spots') which can move with any velocity $v > c$. The same 'spots' can consist also of charged particles, for example, they are formed when suitable particle beams are incident upon a metal plate, and so on. 'Light spots' consist of many particles which are different ones at different times and the particles themselves (photons, electrons, and so on) move with the speed of light or with subluminal speed. Therefore, the existence of these 'light spots' does, in particular, not violate the requirements of causality – it is impoaaible to apply them for superluminal signalization. On the other hand, charged 'light spots' (moving 'patches') in the electrodynamic sense are 'not at all worse' than any other macroscopic charge and in that respect Sommerfeld's calculation is fully applicable (formally this is clear from the fact that the current density $\vec{j} = \rho(\vec{r},t)\vec{v}(\vec{r},t)$, which corresponds to a distributed charge, can move in space with a velocity larger than c). Optical 'light-spots' (that is, 'light spots' formed by photons) which move with a velocity $v > c$ can also emit, but in this case one must take nonlinear effects into account. Briefly, radiation sources moving with velocities larger than c are fully realizable and there is thus a possibility to observe the Cherenkov effect both in vacuo (it is true, under normal conditions only when there are boundaries present) and also in an isotropic plasma.

The Cherenkov effect is possible in vacuo also far from any boundaries, but when there is a strong constant magnetic field H present which is comparable with the well known critical value

$$H_c = m^2c^2/e\hbar = 4.4 \times 10^{13} \text{ Gauss}$$

(here e and m are the electron charge and mass). It was noted already in the beginning of the thirties that in a strong field vacuum behaves like a birefringent medium. In some cases the refractive index for weak electromagnetic waves propagating in a strong magnetic field $n_i > 1$ and, hence, a uniformly moving charged particle can emit Cherenkov waves. We must here not be confused by the fact that (unless we consider motion strictly in the direction of the strong magnetic field) a charged particle can be deflected by the magnetic field. The fact is that we can, in principle, impose a constant velocity v on the particle by some external means (sources). Moreover, we can, formally, assume the mass m of the particle to be arbitrarily large – its velocity will then also be constant.

I wish to make yet one more important methodological remark. Tamm and Frank obtained formula (6) by evaluating the electromagnetic energy flux S through a cylindrical surface surrounding the trajectory of the particle. However, I obtained (of course, later; in fact, in 1939) the same formula by evaluating the change in the energy dW/dt of the electromagnetic field per unit time in the whole of space. Finally, the same result can be obtained (if I am not mistaken this was done first by E. Fermi and afterwards by L.D. Landau) by evaluating the work done by the field on the particle per unit time, that is, the quantity $e(\vec{E}\cdot\vec{v})$, where $\vec{E}$ is calculated at the position of the charge (in the sense, clearly, that $e\vec{E}$ is the radiative friction force: the other parts of the field do not contribute to the relevant expression). One should perhaps expect from the very beginning that all three methods would give identical results. Indeed, this is the case for the Cherenkov effect. However, in the general case for non-stationary charges in vacuo (and in a medium) and, for example, for the case of the transition radiation the quantities S, dW/dt, and $e(\vec{E}\cdot\vec{v})$ are, generally speaking, different; and we must not forget this (for details see Refs. 1,2).

III. THE QUANTUM THEORY OF THE CHERENKOV EFFECT

Whole books have been written about the Cherenkov and related effects, as we have already mentioned, and one might well write yet another one. However, this is now not my problem. I shall therefore dwell only upon one problem relating to the Cherenkov effect, namely, how it can be treated quantally. Generally speaking the Cherenkov effect is, as a rule, completely satisfactorily described in the framework of classical theory and the corresponding quantum corrections are not important. However, it seems to me that from a methodological (and, if you like, a physical) point of view the quantum approach is useful and interesting. However, I am perhaps influenced by the fact that this problem was considered in one of my first papers, published in 1940. When at that time L.D. Landau heard about that paper, he did not think it interesting. In general, let me use this case to note (on the whole this is understandable and well known) that not only in arts, but also in science preferences and tastes are very varied and different. For instance, I literally love problems connected with the Cherenkov radiation, and so on. At the same time, Landau, probably not by chance, considered the quantum theory of the Cherenkov effect uninteresting as he in general did not have any special interest in this effect, probably, one could say, he did not consider it beautiful. This is not at all a criticism, but simply a statement. Its aim was only to emphasize that, in my opinion, there are no grounds to banish from the scientific literature, as is often done, everything personal, and to aim only at a dry statement of facts and description of formulae. Besides, much depends also on the genre of the literature – it is one thing for textbooks, another for popular papers or lectures.

Let us turn, however, to the quantum theory of the Cherenkov effect and let us restrict ourselves to obtain the condition (4) for emission and its quantum generalization (of course, quantum theory enables us also to obtain formula (6) with its quantum corrections).

How can one quantum-mechanically explain the absence of emission by a charge or another static source which moves uniformly in vacuo? To do this it is sufficient to use the energy and momentum conservation laws:

$$E_0 = E_1 + \hbar\omega \ , \quad E_{0,1} = \sqrt{(m^2c^4 + c^2p^2_{0,1})} \ , \tag{9}$$

$$\vec{p}_0 = \vec{p}_1 + \hbar\vec{k} \ , \quad \hbar k = \hbar\omega/c \ , \quad \vec{p}_{0,1} = \frac{m\vec{v}_{0,1}}{\sqrt{[1-(v_{0,1}/c)^2]}} \ , \tag{10}$$

where $E_{0,1}$ and $\vec{p}_{0,1}$ are the energy and momentum of the charge (source) of mass m before (0) and after (1) the emission of a photon of energy $\hbar\omega$ and momentum $\hbar\vec{k} = (\hbar\omega/c)(\vec{k}/k)$ (ω is the frequency of the radiation). One can verify that it is impossible (and this is also clear from the formula (13) with $n = 1$, which follows below) to satisfy the relations (9),(10) for $\omega > 0$, that is, emission is impossible.

In order to consider the problem of the emission by a source in a medium one must, clearly, know only one thing – what are in that case the energy and momentum of the radiation (the expression for the energy $E_{0,1}$ of the source is, clearly, not changed). It is not that simple to do this fully consistently, but on an intuitive level the answer is clear at once. Indeed, the presence of a medium which does not move or change with time does not affect at all the frequency ω but the wavelength in the medium is $\lambda = \lambda_0/n(\omega)$, where $\lambda_0 = 2\pi c/\omega$ is the wavelength in vacuo; in other words, in the medium the wavenumber is $k = 2\pi/\lambda = \omega n(\omega)/c$. If we agree with that remark, we must, instead of (10) put

$$\vec{p}_0 = \vec{p}_1 + \hbar\vec{k} \ , \quad k = \frac{\hbar\omega n(\omega)}{c} \ , \quad \vec{p}_{0,1} = \frac{m\vec{v}_{0,1}}{\sqrt{[1-(v^2_{0,1}/c^2)]}} \ . \tag{11}$$

The simultaneous solution of equations (9) and (11) leads to the result

$$\cos\theta_0 = \frac{c}{n(\omega)v_0}\left(1 + \frac{\hbar\omega(n^2-1)}{2mc^2}\bigg/\sqrt{\left[1-\frac{v^2}{c^2}\right]}\right), \tag{12}$$

or

$$\hbar\omega = \frac{2\frac{mc}{n}\left(v_0\cos\theta_0 - \frac{c}{n}\right)}{\left(1-\frac{1}{n^2}\right)\bigg/\sqrt{\left[1-\frac{v_0^2}{c^2}\right]}} \ , \tag{13}$$

where θ_0 is the angle between $\vec{v}_0$ and $\vec{k}$. If

$$\frac{\hbar\omega}{mc^2} \ll 1 \ , \tag{14}$$

(or a somewhat more general inequality which can be found from (12)), which corresponds to the classical limit, expression (12) changes to (4), as one should have expected. The classical limit corresponds, clearly, to neglecting the recoil which occurs when a 'photon in the medium' with momentum $\hbar\vec{k}$

is emitted. It is also clear from (13) that $\omega > 0$ and $\cos\theta_0 < 1$ that is, emission is possible, only if $v_0 > c/n(\omega)$ (see (5)). In the classical limit when the result (in this case expression (4)) does not contain the quantum constant $\hbar$, the quantum theoretical calculation has merely a methodological character; it may turn out to be convenient, but it is not obligatory: of course, the energy and momentum conservation laws can be formulated also in the classical region. We need solely take into account the connection between the emitted energy W and the change in the momentum of the radiation and of the medium $\vec{G}$. In accordance with (11) we must put

$$\vec{G} = \frac{Wn}{c}\frac{\vec{k}}{k} \tag{15}$$

Further, for a freely moving particle with sufficiently small changes in in energy and momentum,

$$\Delta E = (\vec{v}\cdot\Delta\vec{p}) \equiv (\vec{v}\cdot[\vec{p}_1 - \vec{p}_0])$$

(indeed

$$dE/d\vec{p} = (d/d\vec{p})\sqrt{(m^2c^4 + c^2p^2)} = c^2\vec{p}/E = \vec{v}),$$

we can put $v_0 \approx v_1 \approx v$ and from the conservation laws (9) and (11) and replacing $\hbar\omega$ by W we get

$$\Delta E = W = (\vec{v}\cdot\Delta\vec{p}) = (Wn/c)([\vec{k}/k]\cdot\vec{v}).$$

It is clear that the energy W drops out and we get at once the classical condition for emission $(nv/c)\cos\theta = 1$ (see (4)).

We must still add that the relation (15) or $k = \hbar\omega n/c$ (see (11)) corresponds to writing the energy-momentum tensor in a medium in the Minkovskii form. In actual fact, however, the energy-momentum tensor of the field in the medium has the form proposed by Abraham (we have in mind the simplest case of a non-dispersive medium) which is reflected in the existence of the Abraham force which acts upon the medium with a density

$$\vec{f}^{\mathbf{A}} = [(n^2-1)/4\pi c](\partial/\partial t)[\vec{E}\wedge\vec{H}].$$

One can easily show, however, that expression (15) is valid also when we use the Abraham tensor, if we are interested in the total momentum of both the radiation and the medium (for details see Ref. 1, Chapter 12). However, it is just that quantity which occurs in the conservation law (11) and its classical counterpart. Incidentally, one can say that the necessity to use just expressions (11) and (15) for $\hbar\vec{k}$ or $\vec{G}$ is clear already from the obvious correctness of the result obtained — the classical formula (4) for the angle of the Cherenkov radiation.

IV. THE QUANTUM THEORY OF THE DOPPLER EFFECT IN A MEDIUM

The quantum theory of the Cherenkov effect in the classical limit (14) does not give anything new, apart from an understanding of the rôle played by the conservation laws, and so on. It is therefore rather interesting that in more complicated cases quantum theory enables us to reveal interesting points even in the classical limit. To illustrate this we consider the Doppler effect in a medium.

We remind ourselves first of all of the classical situation using as an example an oscillator with eigenfrequency ω_{00} (this is the frequency in a frame of reference in which the oscillator as a whole is at rest). If the oscillator moves in vacuo with a constant velocity $\vec{v}$ (in the laboratory frame of reference), in this laboratory frame the frequency of the waves emitted by it is equal to

$$\omega(\theta) = \frac{\omega_{00}\sqrt{[1-(v^2/c^2)]}}{1-\frac{v}{c}\cos\theta} = \frac{\omega_0}{1-\frac{v}{c}\cos\theta}, \tag{16}$$

where θ is the angle between the wavevector $\vec{k}$ (direction of observation) and $\vec{v}$; ω_0 in (16) is the frequency of the oscillator in the laboratory frame.

Let there now be a transparent medium (with index of refraction $n(\omega)$) which is at rest in the same frame as the laboratory frame considered. Incidentally, we should not be confused by the fact that the motion of the source in a medium may be accompanied by large energy losses and, importantly, by the destruction of the source itself (say, an excited atoms). The fact is that in a medium one can make an empty gap or channel with dimensions much smaller that the wavelength of the radiation in which we are interested. The energy losses are then steeply reduced and the emission is little changed or, at any rate, is changed in a way which can easily be taken into account (see Ref. 1, Chapter 7).

When there is a medium present formula (16) is replaced by the following one:

$$\omega(\theta) = \frac{\omega_{00}\sqrt{[1-(v^2/c^2)}}{\left|1-\frac{v}{c}\,n(\omega)\cos\theta\right|} = \frac{\omega_0}{\left|1-\frac{v}{c}\,n(\omega)\cos\theta\right|}. \tag{17}$$

One can obtain (17) from (16) by using the general rule – replace v/c by $v/c_{ph} = vn/c$ (in the expression $\sqrt{[1-(v^2/c^2)]}$ one should, of course, not make this substitution, as it does not refer to the emission process). However, one can, of course, also obtain formula (17) automatically by solving the problem of an emitter moving in a medium. Absolute values occur to some extent non-trivially in (17). Of course, if the motion is subluminal ($v < c/n$) or

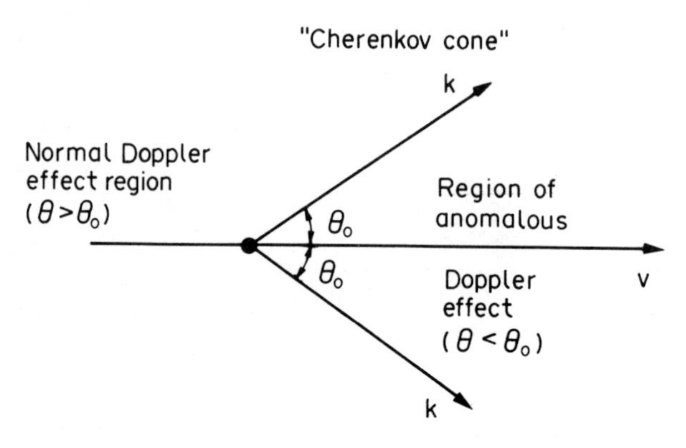

Fig. 4.2

when the emission is, for a superluminal motion, outside the 'Cherenkov cone', that is, when

$$\frac{v}{c}\, n(\omega) \cos\theta < 1 \,, \tag{18}$$

we are dealing with the ordinary, normal Doppler effect; it is true that also in that case the so-called complex Doppler effect, which is caused by dispersion, that is, the ω-dependence of n is possible (Frank, in 1942, was the first to consider the problem of the complex Doppler effect as well as formula (17) with absolute values). In the present case we can use 'Cherenkov cone' to indicate the critical direction of $\vec{k}$, rather than an actual direction in ordinary space (compare Figs. 4.1 and 4.2).

If, however, the motion is superluminal, when the condition

$$\frac{v}{c}\, n(\omega) \cos\theta > 1 \tag{19}$$

holds, that is, where there is emission inside the 'Cherenkov cone' (Fig. 4.2) formula (17) without the absolute signs would lead to negative values of the frequency ω. From that it is already clear that it is necessary to introduce the absolute signs, but we can check this also by other means (in particular, it follows from the calculations given in what follows).

In the region (19) we speak of the anomalous Doppler effect. When dispersion is taken into account the whole picture is rather complicated, but we are here interested in another side of the question and we shall therefore neglect dispersion. In that case it follows from (17) with $n(\omega) = n =$ constant that on the Cherenkov cone itself (when $(v/c)\, n \cos\theta_0 \equiv (v/c)\, n \cos\theta_0 = 1$; see (4)) the frequency $\omega(\theta_0) = \infty$, while $\omega(\theta) \to \infty$ as $\theta \to \theta_0$ on both sides of the cone. It is impossible to say anything more on the basis of formula (17) and the difference between the normal and the anomalous Doppler effect is not particular profound.

We now turn to a quantal derivation of the formula for the Doppler effect in a medium. To do this we must, of course, use the conservation laws (9) and (11), but replacing the particle energies $E_{0,1}$ for the charge (or, more generally, for a source without internal degrees of freedom) by the expression

$$E_{0,1} = \sqrt{[(m+m_{0,1})^2 c^4 + c^2 p_{0,1}^2]},$$

where

$$(m+m_0)c^2 = mc^2 + w_0$$

is the total energy of the system (atom) in the lower state 0 and

$$(m+m_1)c^2 = mc^2 + w_1$$

the same energy in the upper state; we have here for the sake of simplicity considered a system with two levels or, simply, discussed a well defined transition in an atom where we have called the state with the larger energy the upper level (that is, $w_1 > w_0$ and the frequency of the radiation by the atom at rest is $\omega_{00} = (w_1 - w_0)/\hbar > 0$).

If we now apply the conservation laws in the classical limit (14) and use the relation

$$\Delta E = E_1 - E_0 = (\vec{v} \cdot \Delta\vec{p})$$

we find formula (17). However, when we track down the signs (this is simple algebra and we shall not go into details) we can observe an important fact which is, of course, completely hidden in the classical derivation of formula (17). Namely, in the normal Doppler effect region (18) the emission at frequency ω corresponds to a transition of the atom from the upper state 1 to the lower state 0 (the direction of the transition is determined from the requirement that the energy of the emitted quantum $\hbar\omega$ is positive, that is, from the requirement $\omega > 0$). We are used to that situation being the only one and only it occurs naturally in vacuo. However, in the anomalous Doppler effect region (19), that is, when a quantum of energy $\hbar\omega > 0$ is emitted inside the Cherenkov cone, the atoms must make a transition from below (state 0) upwards (to the state 1). Of course, there is here no contradiction — the energy which goes into exciting the emitting system (the atom), and also the energy $\hbar\omega$ of the radiation itself derives from the kinetic energy of the translational motion. Therefore, for the case of superluminal motion $(v > c/n)$ when only the anomalous Doppler effect is possible, even if the emitting atom is initially not yet excited (it starts from the lower energy level 0), it will be excited with the simultaneous emission of quanta inside the Cherenkov cone. Then, an excited atom emits, making a transition into the lower state while radiating quanta at angles $\theta > \theta_0$, that is, outside the Cherenkov cone.

I feel that it would have been rather difficult to suspect this unusual picture without a quantummechanical calculation (however, it is clearly not the quantum effects but the use of the energy and momentum conservation laws which are important). However, if we know what we have just found we can confirm the result and proceed from there using the classical calculations of the radiative friction force which acts upon a superluminal oscillator. In accordance with the conclusion given here, it is just the waves emitted outside the Cherenkov cone which lead to a damping of the vibrations of the oscillator; on the other hand, waves emitted inside the cone (the anomalous Doppler effect) will pump the vibrations of the oscillator (see Ref. 1, Chapter 7).

V. TRANSITION RADIATION

We have seen that when a source (charge, multipole) moves uniformly and rectilinearly in a medium with velocity $v > c/n(\omega)$ Cherenkov radiation occurs. In this case Cherenkov radiation is the only possibility, if the medium is uniform in space and does not change with time.

If, however, the medium is not uniform and/or changes in time or when such a medium lies near the trajectory of a source, the situation is considerably altered. In general, under such conditions it is just transition radiation which appears, in the broad sense of the word, that is, the radiation occurring when a charge (or other sources which do not possess an eigenfrequency) move uniformly and rectilinearly under non-uniform conditions — in a non-uniform medium, in a medium which changes with time, or near such a medium. Of course, transition radiation can in the general case co-exist and interfere with the Cherenkov radiation and with the radiation arising due to the acceleration of the charge (that is, with bremsstrahlung, synchrotron radiation, and so on). However, for an understanding of the physics of the case it is natural, at any rate initially, to separate off just the transition radiation.

We therefore consider a charge moving with a constant velocity*

$$v < \frac{c}{n}$$

when there is no Cherenkov radiation. If we, moreover, consider the vacuum

---

* When there is (Cherenkov, transition) radiation present the energy of the charge changes, in general, and in this connection one meets with the question: Can we assume the velocity of the charge to be strictly constant? The answer is here unconditionally positive for the reason given earlier in section 2. It is a different matter that in some problems one must take into account also the change in the velocity of the source, but that is altogether another question.

($n = 1$) there is, in general, no radiation whatever. In order that it occurs in vacuo the charge (or multipole) must be accelerated or, in other words, the parameter v/c which characterizes the radiation must change. When there is a transparent medium present this parameter already has the form $v/c_{ph} = vn/c$ – it is equal to the ratio of the particle velocity v to the phase velocity $c_{ph} = c/n(\omega)$.

However, and this is the crux of the matter when there is a medium present vn/c can change not only because the velocity v changes, but also due to changes of the phase velocity $c_{ph} = c/n(\omega)$ along the trajectory of the source because of appropriate changes in the refractive index. The radiation which occurs when the parameter vn/c changes due to changes in $n(\omega)$ along or close to the trajectory of a source with v = constant is called transition radiation. To be precise, in the general case of an absorbing medium $\sqrt{\varepsilon} = n + i\kappa$ plays the rôle of the refractive index, where ε is the complex dielectric permittivity of the medium.

The simplest problem of this kind is when a charges passes through the dividing boundary between two media (or, in particular, between vacuum and a medium). I.M. Frank and I considered in 1944 just the simplest kind of transition radiation which occurs in this case. One can note that in some sense transition radiation is a simpler effect than Cherenkov radiation. The fact that the possibility for the occurrence of transition radiation was elucidated with such a great delay is, apparently, explained by the same reason as in the case of the Cherenkov effect.

We note that the explanation given above of the nature of the transition radiation is connected with the fact that it appears when vn/c changes, it is altogether rather formal and, in fact, requires an understanding of the theory of radiation in a medium. It therefore is probably not superfluous to mention the most intuitive explanation of the cause of the appearance of transition radiation when a charge passes through a dividing boundary. It is well known that the electromagnetic field in the first medium (in the medium in which the charge moves at the time considered) can be represented as the field of the charge itself and the field of the 'mirror image charge' which moves in the second medium towards the charge. When the charge and its image passes through the boundary 'from the point of view' of the first medium they, as it were, 'annihilate' or 'reorganize' themselves and this leads to the radiation. Particularly simple is, of course, the case of normal incidence of the charge from vacuum on a perfect mirror – when the charge and its image traverse the boundary of the mirror they 'annihilate' one another completely or, one can say rather, they stop at the boundary (in the sense that the

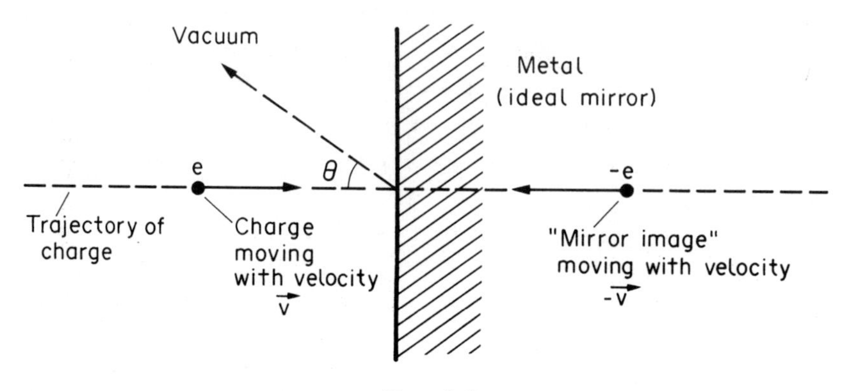

Fig. 4.3

radiation which occurs in vacuo is the same as the radiation by an incident charge e and its image $-e$ which simultaneously stop at the dividing boundary: Fig. 4.3).

To find the emitted energy W in this simplest case we do not need to solve the boundary problem which in general is rather unwieldy, but we can use a rather simple and well known formula for the radiation by charges which suddenly change their velocity*

$$W(\omega,\theta,\phi) = \frac{1}{4\pi^2 c^3}\left\{\sum_i e_i \frac{[\vec{v}_{i2}\wedge\vec{s}]}{1-\frac{(\vec{s}\cdot\vec{v}_{i2})}{c}} - \frac{[\vec{v}_{i1}\wedge\vec{s}]}{1-\frac{(\vec{s}\cdot\vec{v}_{i1})}{c}}\right\}^2 , \tag{21}$$

where e_i is the charge of the ith particle, the velocity of which changes suddenly from a value $\vec{v}_{i1}$ to a value $\vec{v}_{i2}$, while $\vec{s}=\vec{k}/k$ is the direction of the wavevector of the radiation which is characterized by the angles θ and ϕ; the total frequency-dependent energy of the emitted radiation is

$$W(\omega) = \int W(\omega,\theta,\phi)\sin\theta\, d\theta\, d\phi ,$$

and the total energy is $W=\int W(\omega)\, d\omega$.

If in vacuo (in the absence of boundaries and so on) a single charge $e_1=e$ suddenly stops or suddenly accelerates from a state of rest to a velocity v we have (see (21))

---

* The sudden change in the velocity means that it occurs during a time τ which is short compared to the period $T=2\pi/\omega$ of the emitted wave which we are considering (to be more precise, $\tau \ll T = 2\pi/\omega$ refers only to non-relativistic motion; in the general case, it is necessary that the time τ be small compared to the formation time $t_f = 2\pi\,[\omega\{1-(v/c)\cos\theta\}]$, which is introduced below (see p.120), for the case $\theta \gtrsim 0$, where θ is the angle between the velocity $\vec{v}$ and the wavevector $\vec{k}$; $k=\omega/c$, when we consider emission in vacuo).

$$W(\omega,\theta,\phi) = \frac{e^2 v^2}{4\pi^2 c^2} \frac{\sin^2\theta}{(1-\frac{v}{c}\cos\theta)^2}, \quad W(\omega) = \frac{e^2}{\pi c}\left(\frac{1}{v/c}\ln\frac{1+\frac{v}{c}}{1-\frac{v}{c}} - 2\right). \quad (22)$$

In the case of transition radiation on a perfect mirror one must assume that the charge $e_1 = e$ with velocity $\vec{v}$ and the charge $e_2 = -e$ with velocity $-\vec{v}$ suddenly stop at the boundary (see Fig. 4.3). As a result (after integrating over ϕ which means multiplying by 2π) one will observe in medium 1 (in vacuo) radiation with an energy

$$\left.\begin{aligned} W_1(\omega,\theta) &= \frac{e^2 v^2 \sin^2\theta}{\pi^2 c^3 [1-(v/c)^2\cos^2\theta]^2}, \\ W_1(\omega) &= \frac{e^2}{\pi c}\left[\frac{1+\left(\frac{v}{c}\right)^2}{2v/c}\ln\frac{1+\frac{v}{c}}{1-\frac{v}{c}} - 1\right]. \end{aligned}\right\} \quad (23)$$

Here θ is the angle between $\vec{k}$ and $-\vec{v}$, shown in Fig. 4.3. In the non-relativistic case (that is, when $v \ll c$)

$$W_1(\omega,\theta) = \frac{e^2 v^2 \sin^2\theta}{\pi^2 c^3}, \quad W_1(\omega) = \frac{4e^2 v^2}{3\pi c^3}. \quad (24)$$

In the ultrarelativistic limit $(v \to c)$ we have

$$W_1(\omega) = \frac{e^2}{\pi c}\ln\frac{2}{1-\frac{v}{c}} = 2\frac{e^2}{\pi c}\ln\frac{2E}{mc^2}, \quad E = \frac{mc^2}{\sqrt{[1-(v/c)^2]}}, \quad (25)$$

which is the same as the radiation of a single particle in this limit (see (22)); this is understandable as for $v \to c$ the radiation is directed along the particle velocity, as a consequence of which the radiation of the charge e 'entering the metal' is not observed (in the medium 1, the vacuum, only the radiation from the mirror image — the charge $-e$, which has a velocity $-\vec{v}$ — is observed). In the non-relativistic limit the energy (24) is four times larger than the energy (22) for a single charge radiating into the semi-sphere of directions (into the vacuum), since for a non-relativistic velocity the field of the charge e and that of its mirror image $-e$ add, that is double.

It is absolutely obvious that the transition radiation discussed here occurs when boundaries of any media with different 'electrical' parameters (such as the dielectric permittivity or the refractive index) are crossed. However, initially the attention was concentrated on the incidence of a charge upon a metal (which might not be a perfect mirror) and, hence, on transition radiation — in the main optical — in the 'backward' direction, which is observable in the vacuum. For relativistic particles however, a different version in which a particle passes through a medium and emerges into the vacuum is fully realistic for relativistic particles with a sufficiently high energy. From a purely theoretical point of view this problem is practically equivalent to the previous one and the corresponding formula for the intensity of the radiation is obtained simply by replacing the velocity v by the velocity $-v$ (*vide infra*). Of course, at the same time there is no symmetry when we evaluate the field in these cases and, in general, the intensities are different when we replace v by $-v$ (or, in other words, when the particle enters or leaves the medium, or for, respectively, 'backward' and 'forward' observations) and in some cases the differences will be large. For the forward radiation and, in particular, when the particle leaves the medium and enters the vacuum there are higher frequencies in the radiation spectrum and, to be concrete, in a condensed medium the transition radiation by relativistic particles can stretch out into the X-ray part of the spectrum. To give here the solution of the corresponding boundary problems is not expedient (see Ref. 2 and the literature quoted there). It is appropriate, however, to give a final formula for the case where the medium 1 is the vacuum and the medium 2 is described by a complex permittivity ε:

$$W_1 = \frac{e^2 v^2 \sin^2\theta \cos^2\theta \left|(\varepsilon - 1)[1-(v^2/c^2) + (v/c)\sqrt{(\varepsilon - \sin^2\theta)}]\right|^2}{\pi^2 c^2 \left|[1-(v/c)\ \cos^2\theta][1+(v/c)\sqrt{(\varepsilon - \sin^2\theta)}][\varepsilon\cos\theta + \sqrt{(\varepsilon - \sin^2\theta)}]\right|^2} . \tag{26}$$

For a perfect mirror we may assume that $|\varepsilon| \to \infty$ and expression (26) goes over into (23), as should be the case. The expression for W_2 referring to the case where the charge e leaves a medium of permittivity ε (medium 1) with a velocity $\vec{v}$ and enters the vacuum (medium 2) is obtained from (26) by replacing v by $-v$; moreover, the angle θ is now the angle between $\vec{k}$ and $\vec{v}$ (and not between $\vec{k}$ and $-\vec{v}$, as in (26)). Replacing v by $-v$ in (26) is not altogether harmless — in the denominator there appears a factor

$$[1-(v/c)\sqrt{(\varepsilon - \sin^2\theta)}]$$

instead of

$$[1+(v/c)\sqrt{(\varepsilon - \sin^2\theta)}] .$$

This is just the reason why, as we mentioned, there appear higher frequencies when a particle leaves the medium (we must take here into account that ε approaches unity for high frequencies). As a result the total intensity (integrated over angles and frequencies) also increases; in the simplest case it turns out to be proportional to

$$[1-(v^2/c^2)]^{-\frac{1}{2}} = E/mc^2$$

(E is the total mass of the emitting particle of mass* m). This important fact was made clear only in 1959. This opened up much wider perspectives for the creation of efficient 'transition counters' intended for the detection or, more exactly, for the velocity determination of relativistic particles. It is true that the problem of transition counters which is connected with having to consider many dividing boundaries had been raised already earlier but the proposed use of transition radiation in the optical region of the spectrum can, apparently, not lead to our goal.

As often happens, the appearance of a possibility for a 'practical' application, in this case to high energy physics, suddenly increased the interest in transition radiation. In fact, according to a bibliographical index I have, from 1945/6 up to 1958 inclusive, fourteen papers were devoted to transition radiation, in the next thirteen years (1959 to 1971) another 244 papers appeared, but between 1972 and 1978 there appeared 296 papers (the last year has not been taken into account completely). We note also that the first published experimental results, in which transition radiation was observed, refer to 1959. Altogether, transition radiation, this rather simple and clear effect from the field of classical electrodynamics, attracted hardly any attention to it for about fifteen years, but now it is very popular, although mainly in the context of developing and applying transition counters. This side of the problem has been mentioned even in a number of survey articles, apart from a large number of papers (for references see Ref. 2).

Without wanting to deny to any extent the importance of the investigations connected with transition counters I wish to emphasize that transition radiation, taken in a rather broad sense, is undoubtedly of value from a general point of view, it is connected with some definite ideas and has its own 'language' and thereby facilitates further developments in a number of directions. The situation is here, in general, similar to the one with regards to the Cherenkov effect which is directly (and one could say in a pure form) used first of all in Cherenkov counters.

---

* The fact that for an ultrarelativistic particle ($v \to c$) the total energy emitted in the forward direction (say, when the particle emerges from the medium into the vacuum) is appreciably larger than the total energy emitted in the backward direction (for instance, into the vacuum) may nevertheless cause some surprise. One must bear in mind that for an ideal mirror and when $v \to c$ the energies emitted forwards and backwards are equal to one another. However, when one takes into account the frequency dispersion, especially at high frequencies, when the ideal mirror approximation is totally inapplicable, as $v \to c$ it is not the charge e itself which emits the backwards radiation, but its appreciably weaker mirror image. On the other hand, the charge e itself emits in the forward direction. That is, in final reckoning, the reason why the total emission in the forwards direction dominates as $v \to c$.

Our attention has so far been concentrated on the transition radiation which occurs when a single particle crosses the dividing boundary or a number of boundaries between media. In the last case we are dealing with either an ordered succession of boundaries, that is, a system with a well defined period, or with randomly distributed boundaries (inhomogeneities).* Another, long standing trend is determined by the fact that any radiation and, in particular, transition radiation with a wavelength λ is produced not in a point but in some region ('the formation zone') with dimensions which are determined by the wavelength λ but which can also be appreciably larger than it. As we have already mentioned, this is the reason why the Cherenkov effect occurs also when a particle moves in vacuo but close to a medium (in a channel, in a gap, or close to the boundaries of a medium). Completely analogously transition radiation (which in this case is usually called diffraction radiation) occurs when a source (charge) moving uniformly in vacuo (or in a uniform medium) passes close to some obstacles — metallic or dielectric globules, diaphragms, a diffraction grating, and so on. Neglecting the remarks made earlier which were of a general nature, the occurrence of this kind of transition radiation can also be easily explained using the method of images.

In the case of relativistic particles when we consider the radiation along the direction of their velocity the formation zone, in general, increases strongly with increasing energy. For instance, in vacuo the size of the formation zone L_f in the direction of the velocity for a given radiation wavelength increases proportional to $(E/mc^2)^2$, where E is the total energy of the charge (source) and m is its rest mass (we assume that $E \gg mc^2$; in the general case in vacuo in the direction of the velocity

$$L_f = vt_f = (\lambda v/c)[1-(v/c)]^{-1} ,$$

where λ is the radiation wavelength).† Under conditions when the size of the formation zone L_f is large, or the time for the formation $t_f = L_f/v$ is

---

* Transition radiation in a periodically inhomogeneous medium has, of course, its own characteristics. This is, perhaps, the reason why it is sometimes called resonance radiation. It seems to me, however, that to use different names for different forms of transition radiation is hardly advisable and can only lead to confusion. I shall call transition radiation in a periodic medium resonance transition radiation and also transition scattering (the reason for the last name will become clear later on).

† After the time of formation $t_f = L_f/v$ the light (radiation) is ahead of the particle by a wavelength λ; in fact, $(c-v)t_f = \lambda$; we note that we can also take for the size of the formation zone a length smaller, for instance, by a factor two than the length used by us — we are here talking about a definition of L_f; the quantitative results are, of course, independent of this definition.

large, the effect of the medium along or close to the trajectory of the charge also affects what happens at large distances $L \sim L_f$ and this leads to particular features of the radiation.

One type of transition radiation arises already in a homogeneous medium when its properties change with time. What happens here can be understood more easily if we use a terminology involving the parameter vn/c. We have already mentioned that when v = constant we need for the appearance of transition radiation along the trajectory of the charge or close to it a change in the refractive index n. However, this change occurs, clearly, also under conditions when the index n changes with time, for example when it at some time more or less suddenly increases or decreases.

It is interesting that this kind of transition radiation can occur even for a charge which is at rest relative to the medium. Indeed, if through the application of an (electrical or magnetic) field or by some other means the medium discontinuously (that is, rather suddenly) changes from an optical isotropic to an optical anisotropic medium, the polarization of the medium which surrounds the fixed charge changes and also the spherical symmetry ceases to exist. Such a change in the polarization entails, clearly, the emission of electromagnetic waves.

Like Cherenkov-type radiation transition radiation is very general in character also in the sense that it takes place for various kinds of waves. As an example we mention transition radiation of acoustic waves which arises when a moving dislocation moves through a grain boundary in a polycrystalline body.* It is probable that other problems, connected with transition radiation (see, also section 7 below) are also of interest in acoustics.

Also interesting and with, perhaps, a real value for applications to pulsar magnetospheres is the transition radiation which occurs already in vacuo, but when there is a strong magnetic field present, which leads to the occurrence of nonlinear electrodynamic effects.

---

* More often one considers the emission of sound occurring when a dislocation arrives at a crystal boundary. The closest analogy in that case is not transition radiation, but bremsstrahlung when a charge is stopped. At the same time if we have in mind the radiation and not the 'fate' of the source, transition radiation and bremsstrahlung, with the effect of the boundary taken into account, are in many cases indistinguishable.

VI. TRANSITION SCATTERING. TRANSITION BREMSSTRAHLUNG

If transition radiation takes place when a charge moves in a medium with a periodically (say, sinusoidally) changing refractive index, we can speak not only of transition radiation or resonance transition radiation, but we can also call the corresponding process transition scattering. Indeed, a dielectric permittivity (refractive index) wave, which can be a standing or a travelling wave, is so to speak scattered by a moving charge producing electromagnetic (transition) radiation. However, it would not be good to apply under such conditions the name transition scattering instead of transition radiation unless the effect would remain also in the limiting case of a charge at rest. In that case it would be at least unnatural to speak of transition radiation, while the name 'transition scattering' would reflect the essence of the situation. Indeed, we are, for instance, dealing with the incidence of a permittivity wave on a non-moving (fixed) charge and an electromagnetic wave will go away from the charge.

It is very simple to understand this result apart from any connection with the theory of transition radiation. To see this we consider an isotropic medium which is characterized by a dielectric permittivity ε which depends solely on the density ρ of the medium. If a longitudinal acoustic wave propagates in this medium, the density

$$\rho = \rho^{(0)} + \rho^{(1)} \sin\left[(\vec{k}_0 \cdot \vec{r}) - \omega_0 t\right]$$

and, because of what we have said earlier

$$\varepsilon = \varepsilon^{(0)} + \varepsilon^{(1)} \sin\left[(\vec{k}_0 \cdot \vec{r}) - \omega_0 t\right] , \qquad (27)$$

where $\varepsilon^{(1)}$ is the change in ε caused by the change in ρ (in the simplest case $\varepsilon^{(1)} = \text{constant} \times \rho^{(1)}$); of course, we have chosen a well defined mechanism for the change in ε (in the present case, it is due to a change in the density ρ) only to fix the ideas and to make the discussion easily understandable — only the fact that there is a permittivity wave (27) present in the medium is important for what follows.

We now place in the medium a fixed (or, say, an infinitely heavy) charge e. Around this charge there appears an induction $\vec{D}$ and a field $\vec{E}$:

$$\vec{D}(\vec{r},t) = \varepsilon\vec{E}(\vec{r},t) \quad , \quad \vec{D}^{(0)} = \frac{e\vec{r}}{r^3} \quad , \quad \vec{E}^{(0)} = \frac{e\vec{r}}{\varepsilon^{(0)} r^3} , \qquad (28)$$

where, clearly, the superscript (0) indicates the 'unperturbed problem' — one considers the field of a charge without a permittivity wave. When there is a permittivity wave present, to a first approximation (which corresponds to the assumption that $|\varepsilon^{(1)}| \ll \varepsilon^{(0)}$ which we make here for the sake of simplicity) there arises around the charge a varying polarization

$$\delta\vec{P} = \frac{\delta\vec{D}}{4\pi} = \frac{\varepsilon^{(1)}\vec{E}^{(0)}}{4\pi}\sin[(\vec{k}_0\cdot\vec{r}) - \omega_0 t], \qquad (29)$$

Such a variable polarization which, when $\vec{k}_0 \neq 0$, does not have spherical ssymmetry, but induces, of course, the appearance of an electromagnetic wavee with frequency ω_0 which goes away from the charge (see Fig. 4.4). The wave-number of this wave $k = 2\pi/\lambda = (\omega_0/c)\sqrt{\varepsilon^{(0)}}$. If, as we assumed, the permittivity wave is caused by the acoustic wave, $k \ll k_0 = \omega_0/u$, where u is the sound speed (we assume here, of course, that $u \ll c/\sqrt{\varepsilon^{(0)}}$).

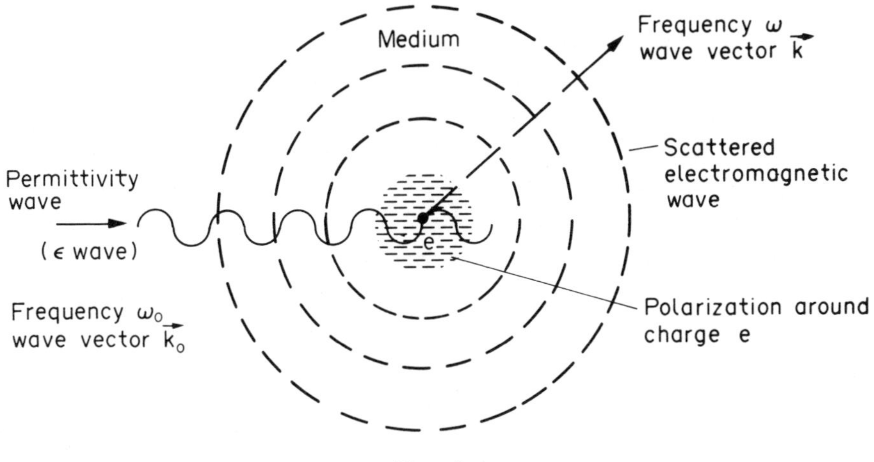

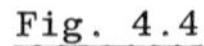
Fig. 4.4

The electromagnetic wave which arises here can be considered to be scattered in the same sense as for other kinds of scattering, such as, for instance, the scattering of an electromagnetic wave by an electron at rest (in this case we have, of course, in mind it being at rest only when we forget about the action of the incident wave). Transition scattering plays an important rôle in plasma physics and is, on the whole, a very general kind of effect which must, for instance, occur in vacuo when an electromagnetic or gravitational wave is incident on a region with a strong constant (static) or quasi-stationary electromagnetic field (an outgoing electromagnetic wave is then produced; see also section 7 below).

Close to the transition scattering process, and related to it, is transition bremsstrahlung. It occurs in a medium if a uniformly and rectilinearly moving charge e passes rather closely to another charge e' which is, for example, at rest. In its characteristics transition bremsstrahlung is similar to ordinary bremsstrahlung although for its occurrence it is not at all necessary that the charge is accelerated (a change in the rectilinear trajectory or a braking of the charge). The term 'transition bremsstrahlung' is justified just because this radiation occurs when particles collide (and radiation produced in collisions is called bremsstrahlung). Moreover, transition bremsstrahlung of electrons is described by expressions which are very close to the formulae for ordinary bremsstrahlung. Furthermore, transition bremsstrahlung interferes with ordinary bremsstrahlung. At the same time, in contrast to ordinary bremsstrahlung, transition bremsstrahlung does, of course, not disappear in the limit of infinitely heavy colliding particles. Somehow or other the general theory of bremsstrahlung of particles in a medium is obliged to take into account also transition bremsstrahlung and its interference with ordinary bremsstrahlung (exactly as the general theory of scattering must take into account transition scattering).

It is easy to understand the physical nature of transition bremsstrahlung if we bear in mind that the field $\vec{E}$ and the polarization $\vec{P}=(\varepsilon-1)\vec{E}/4\pi$ of a uniformly moving charge and, in particular, of a charge at rest can be expanded into waves with wavevector $\vec{k}_0$ and frequency $\omega_0=(\vec{k}_0\cdot\vec{v})$, where $\vec{v}$ is the velocity of the particle. These waves are, in general, connected with permittivity waves with the same values of ω_0 and $\vec{k}_0$. Such permittivity waves 'dragged along' by a single particle undergo scattering by another charge as a result of which electromagnetic radiation is produced — in this case transition bremsstrahlung.

Transition bremsstrahlung may be responsible for the generation of any kind of 'normal waves', which can propagate in the medium considered (excitons, photons, phonons, and so on), when fast particles, dislocations, and so on, pass through the medium. In other words, like transition scattering, transition bremsstrahlung is a phenomenon which has a very general character.

VII. CONCLUDING REMARKS

In the foregoing we have already many times emphasized the very wide value for physics as a whole of the radiation processes occurring when sources move with a constant velocity. With the passing of time ever new problems of this sort arise or attract attention. In this respect the Cherenkov and the Doppler effects are the best known and the most widely studied. However, transition radiation and, especially, transition scattering are so far insufficiently known to all physicists. It is therefore natural, even if we restrict ourselves to electrodynamics that many problems in the theory of transition radiation and scattering remain insufficiently studied. As an example we mention a problem from the field of surface physics: the transition radiation of various surface waves in different situations (when a charge is moving on an inhomogeneous surface consisting of two different media, or passes close to it, and so on), the transition scattering of Rayleigh and other surface waves by charges, defects, and so on, which are situated on or close to the surface, transition radiation and scattering in the case of a rough or regularly-inhomogeneous surface (a lattice, or such like). However, even less studied are the possibilities in acoustics, hydrodynamics, elasticity theory, and so on. For instance, in superfluid helium II, it is well known that several kinds of sound waves can propagate (in the bulk — first and second sound waves, in films — third sound propagates; in helium which occupies a finely porous medium fourth sound can propagate). Several kinds of normal (eigen) waves can propagate in crystals which in the general case are neither purely longitudinal nor purely transverse. Different kinds of sound waves can, in principle, be scattered by various inhomogeneities and perturbations and in the process change into other kinds of waves. I can mention here also transition radiation (creation) of electron-positron pairs which must occur when the dividing boundary of two media or, say, the boundary of an atomic nucleus, is crossed. On the whole, transition radiation and transition scattering are, in principle, possible for any kind of field (particle) with the emission of another kind of field.

To illustrate concretely the usefulness of the idea of transition processes in physics I give an actual example. In 1973 there appeared a paper which within the framework of the general theory of relativity considered a charge situated at the centre of mass of a binary (two identical, electrical neutral stars moving in a circle with respect to their centre of mass). Such a completely fixed charge emits electromagnetic waves. This result is, at first sight, unexpected. However, it is obvious when one knows what transition scattering is, and also bears in mind that in the general theory of relativity the gravitational field affects the electromagnetic properties of the vacuum: one can say that the vacuum possesses an electrical permittivity and a magnetic permeability which depend on the metric tensor g_{ik}. It is thus clear that moving stars modulate the permittivity and this produces permittivity waves. As a result is is natural that transition scattering occurs by the fixed charge and that there appears a 'scattered' electromagnetic wave. This understanding of this fact enables us to consider particularly simply[2] such problems as the transformation (scattering) of a gravitational wave by a charge or, what is more realistic in astrophysics, by a magnetic dipole (pulsar).

An understanding of the nature and features of radiation and scattering processes which occur when sources move uniformly and rectilinearly (Cherenkov and Doppler effects, transition radiation, transition scattering, and related phenomena) can thus be considered to be important and necessary elements of a physical education.

REFERENCES

1. V.L. Ginzburg, *Theoretical Physics and Astrophysics*, Pergamon Press, Oxford, 1979. A second revised Russian edition of this book appeared in 1981, published by Nauka, Moscow; one of the most important additions is a separate chapter devoted to transition radiation and transition scattering.
2. V.L. Ginzburg and V.N. Tsytovich, Phys. Reports, **49** (1979) 1; a few elucidations and corrections of a few errors can be found in Phys. Lett., **79A** (1980) 16.
   A monograph by V.L. Ginzburg and V.N. Tsytovich entitled *Transition Radiation and Transition Scattering (Some Theoretical Problems)* will be published in 1983 by Nauka, Moscow. Plenum Press, New York plans to publish as English translation of this book.
3. O. Heaviside, *Electromagnetic Theory*, The Electrician Publ. Comp., London, 1912, vol.3.

Index

OTHER TITLES IN THE SERIES IN NATURAL PHILOSOPHY